Portland Community College
WITHDRAWN

SolidWorks Basics

A Project Based Approach

Fred Fulkerson

Industrial Press, Inc.

Industrial Press, Inc.

32 Haviland Street, Unit 2C
South Norwalk, Connecticut 06854
Phone: 212-889-6330
Toll-Free in USA: 888-528-7852
Fax: 212-545-8327
E-mail: info@industrialpress.com

SolidWorks Basics: A Project Based Approach
By Fred Fulkerson

ISBN print: 978-0-8311-3593-5
ISBN ePDF: 978-0-8311-9336-2
ISBN ePUB: 978-0-8311-9337-9
ISBN MOBI: 978-0-8311-9338-6

Copyright © 2015 by Industrial Press, Inc.
All rights reserved. Published 2015.
Printed in the United States of America.

This book, or any parts thereof, may not be reproduced, stired in a retrieval system, or transmitted in any form without the permission of the publisher.

Sponsoring Editor: Taisuke Soda
Developmental Editor: Laura Brengelman
Cover Designer: Janet Romano-Murray

Excel® is a registered trademark of the Microsoft Corporation.

LEGO® is a registered trademark of the The LEGO Group.

Mastercam® is a registered trademark of CNC Software, Inc.

SolidWorks® is a registered trademark of Dassault Systèmes SolidWorks Corporation.

Minions™ are characters in the feature films *Despicable Me* (© 2010), *Despicable Me2* (© 2013), and *Minions* (© 2015)*,* which are trademarks of and copyrighted by Universal Studios, LLC. All rights reserved. The LEGO Minion images in this book are original drawings created by the author using SolidWorks and are provided solely as part of the project based instructions designed for teaching SolidWorks.

Instructor's materials for this book can be found at
sites.google.com/site/fredumacations/solidworks-minions

Also visit us at:
industrialpress.com
ebooks.industrialpress.com

Table of Contents

List of Online Instructional Videos..vii
Preface ix
How to Use This Guide...xi

Chapter 1: Getting Started..1
Introduction..1
Starting SolidWorks..1
    User Interface..2
    Tools and Toolbars 7
    File Types...14
    Backing Up Files..14

Chapter 2: Basic Sketching...17
Line/Arc Sketching...17
Fully Defining the Handle 21
Extruding the Handle 24
Editing Your Work 26

Chapter 3: Custom Templates..29
Customizing Your Templates 30
Drawing Templates 31
Toolbox Customization...33
Copy Settings Wizard 35
    Saving Your Settings 35
    Restoring Your Settings...36

Chapter 4: LEGO Minions, Part 1—Designing.................37
2 x 2 Block..37
    Square Base Extrusion...37
    Round Extrusions..40
    Edit a Feature..46
Re-Ordering Items in the Feature Tree 46
    Rollback Bar 47
    Shell 47
    Chamfer..49
Interior Features...50
Wheel Hub 58
Tire..74
Design Configurations—Parametric Modeling 82
    Naming Dimensions..84
    Configuring External Dimensions 85
    Selecting a Configuration 86
    Repairing Issues...87

Creating a New View Orientation..89
Configuring Internal Features 93
Configuring Features 94

Chapter 5: Design Tables with Excel 105
Standard 105
Creating the Design Configurations 107
Bell Type...109
Adding a Feature to the Table...110

Chapter 6: Heat Sink 113
Creating Ribs 113
Extra Practice 125

Chapter 7: LEGO Minions, Part 2—Assemblies 127
Assemblies 127
Inserting a Component 128
Adding Mates 129
Creating a Subassembly within an Assembly 133
Creating Your Minions 134
Copying an Existing Assembly Component 135
Copy with Mates 136
Adding a Part through File Explorer...137
Mirror Assembly Components..139
Assembly Linear Pattern 141
Interference Detection 145
Open Part 146
Inserting a Subassembly 148
Assembly Configurations 150
Pack and Go 152
Exploded Assembly 153

Chapter 8: LEGO Minions, Part 3—Drawing 157
Exploded View...159
Drawing View Property Manager...161
Feature Manager Design Tree 162
Restoring Broken Title Block Link..163
Standard Three View 164
Selecting a Scale 165
Deleting a View 165
Showing Hidden Lines..166
Cross Section 166
Inserting Solid Creation Dimensions 167
Model Items Property Manager 169
Adding Center Mark/Line..171
Adding Text to a Dimension...172
Chamfer Dimension 172

  Dimension Property Manager..175
  Projected View...176
  Inserting Dimensions into One View......................................177
  Smart Dimensioning a Drawing...178
  Broken-Out Section 181
  Detail View 183
  Aligned Section View 184
  Dimensioning to a Virtual Sharp 185
  Creating Drawing Views Using Model View..........................187
  BOM (Bill of Materials) 188
  Showing Part Configurations...191
  Save as a PDF 192

Chapter 9: LEGO Minions, Part 4—Revisions......................193

Chapter 10: Bar Puller..199
Shank..199
Create Remaining Parts..200
  Toolbox 203
Adding a Material 203
  01 Tool Steel Properties..204
Assemblies and Subassemblies...204
  Assembly/Subassembly Creation..205
  Hole Series..207
  Adding Toolbox Components..211
  Edit Toolbox Component..214
Exploded Views..218
  Animating Your Explosion 221
  Motion Study 221

Chapter 11: Bar Puller—Drawings..223
  Break in View...223
  Crop View..224
  Hole Callout 225
  Ordinate Dimensioning..226
  Hole Tables 227

Chapter 12: Centering Jigs—Parts..231
Handle Grip...232
  Cosmetic Thread 232
Handle..233
  Creating a Basic Loft...234
  Hole Wizard..239
  Adding Raised Text 244
  Adding an Appearance..248
Bushings..249

Screw 251
    Sweep Cut Thread 253
    Revolve Cut 257
Spring 258
    Swept Boss 258
    Extruded Cut with a Line 263
Base 267
Center Block 270
    Derived Part 273
Left Jaw 274
Right Jaw—Derived Part 278
    Edit in Context 279
Name Plate 280
    Wrap 281
    Scale 281

Chapter 13: Centering Jig—Assembly 283
Screw Subassembly 283
    Assembly Configurations 284
    Cavity 288
Main Assembly 289
    Flexible Subassembly 290
    Hole Series 290
    Global Variable Creation 292
    In-Context Parts 296
    Face Fillet 304
    Assembly Configurations 307

Chapter 14: Turkey Call 309
Creating the Revolved Part 309
Inserting the Sketch Picture and Creating the Spline 310
    Creating the Spline 312
    New View Orientation 313
Editing the Spline 315
Creating Work Planes 317
    Repairing Sketch Errors 319
Deleting a Face 323

Appendixes 325
Appendix 1: Constraints/Relationships 325
Appendix 2: Dimension Tolerance Types 327
Appendix 3: Short Cuts 328
Appendix 4: Built-In Calculator 329
Appendix 5: New Features of SolidWorks 2015 332

Index 335

About the Author 345

Online Instructional Videos

Show Hide Planes SW 6	CD 00 Creation 106
Docking the Menu Bar 9	CD Naming Dimensions 106
Adding a Command Button 12	CD Design Table 107
Removing a Command Button. 12	CD Bell 110
LA Handle 20	CD Edit Table 111
LA Constraints 22	Heat Sink Rib 117
LA Handle Extrude 25	Rib Linear Pattern 118
LPattern 45	Rollback Bar 119
Re_Arranging Features 47	Create Full Round Fillet 119
Shell 48	Edit LPattern 121
Midpoint Constraint 50	15_5 Plane 122
Extrude to Face 51	Rib Convert Entities 122
Rib Creation 54	Mirror Rib 124
Dimensioning Angles 61	End Rib 124
Diameter Dimensioning 63	Rotate Assembly Component 129
Sketch Mirror 65	Adding a Concentric Mate 130
Switching Sketch Planes 66	Editing a Part in an Assembly 131
Offsetting Sketch Entities 69	Coincident Mate 132
Trimming Entities 70	Creating a Subassembly within an Assembly 133
Circular Sketch Pattern 73	Copying an Assembly Component 135
Symmetrical Relationship 75	Select Other 136
Counter Block Cut Top Slot 77	Copy with Mates 137
SolidWorks Sketch Arch 77	Mirror Assembly Components 140
Share Sketch 81	Evaluate Measure 141
Creating Design Configurations 83	Replacement Component 151
Naming Dimensions 84	Explode Wheel 154
Configure Dimension 85	Radial Explode 156
Deleting Constraints 87	Starting a Drawing 158
Mid Plane Creation 88	Restoring Title Block Links 163
New View Orientation 90	Creating Section Views 167
Copy Sketch 90	Fixing Dimensions 168
Revolve Cut Feature 98	Editing Dimension Text 172
Create Full Round Fillet 104	

Adding a Chamfer Dimension 173	Extrude up to Body 238
Setting Dimension Precision 173	Hole Wizard 241
Editing Your Drawing Notes 174	Adding Raised Text 248
Creating a Projected View 176	Adding a Cast Iron Appearance 248
Breaking the View Alignment 177	Creating a Helix 254
Broken-out Section 181	Thread-Swept Cut 256
Adding a Linked Note 182	Sweep Cut Thread 256
Creating a Detail View 183	Derived Sketch 259
Creating an Aligned Section 184	Composite Curve 261
Dimensioning to a Virtual Sharp 185	Sweep Boss 262
Reordering Drawing Sheets 187	Cutting with a Line 263
Creating a Bill of Materials 188	Adding a Simple Equation 265
Adding Balloon Items 189	Insert into New Part 273
Magnetic Line 190	Creating a Derived Part 279
BP Shank Sketch 200	Editing a Derived Part 279
BP Hole Wizard 200	Inserting a Sketch Picture 281
SW New Material 203	Wrap Feature 281
BP Sub Assembly 205	Creating a Limit Distance Mate 286
BP Insert Component 207	Symmetry Mate 287
BP Hole Series 210	Screw Mate 288
BP Insert Fastener 212	Using Cavity to Create Internal Threads 288
BP Assembly Main 213	Creating Global Variables 292
BP Copy Parts 215	Creating an In Context Part 297
Change Transparency 216	Adding in Context Hole Wizard Holes 299
Mirror Subassembly to Create Opposite Hand Version 217	In Context Nylon Bushing 302
Threaded Explosion 220	Adding a Folder for Fasteners 302
Break in View 224	Up to Vertex Extrusion 303
Creating a Cropped View 225	Face Fillet 304
Creating a Hole Callout 225	TC Sketch Picture 312
Creating Ordinate Dimensions 227	TC Spline 313
Creating a Hole Table 228	TC Create View 314
Creating Cosmetic Threads 233	Deleting Constraints 321
Creating a Basic Loft 236	TC Delete Face 323

Preface

This book is for anyone who wishes to learn SolidWorks and has little or no prior knowledge of any CAD software. It also is for anyone who uses SolidWorks to design and prototype parts, jigs, fixtures, and manufactured components or for 3D visualization, since the contents teach both specific SolidWorks steps and good design skills.

All of the instructions are project based and focus on the development of sound modeling skills. Parts are created using most of the available solid modeling features of the software, which are brought together in fully functioning assemblies and result in the creation of part and assembly drawings.

The LEGO™ Minion™ projects were designed to be fun and productive exercises. The rest of the projects not only show how to take advantage of SolidWorks' many features but also teach tolerancing and some basic tool design, while stressing the importance of a solid design intent.

Like any good software package, SolidWorks provides multiple ways to arrive at your final design. This book will lead you down many of these pathways, resulting in a large resource base to draw from when encountering new design challenges.

One of the nicest aspects of SolidWorks is that it is such an intuitive design tool, making it easy to learn the basics. Yet it is so complex and capable of such a wide range of tasks that once you start learning it, you will enjoy the challenge of doing more with this simple but advanced software tool. After completing this book, you will have a solid foundation on which to continue building your skills.

Acknowledgments

I would like to thank the following teachers/professors for providing valuable input in the development of this book: Henry Kastner, Mike Verwey, Bradley Smith, Don Yusep, and Bill Ferguson.

I also would like to thank my employer, Conestoga College Institute of Technology and Advanced Learning, and particularly Mitchell Wawzonek, Chair of Enginneering and Information Technology, for encouraging me to undertake this project.

Thanks also to my students who made numerous suggestions on how the LEGO Minion material could be clarified.

Finally, I would like to thank my family for all of their support while I toiled away in the basement.

How to Use This Guide

To best address various learning styles, the exercises in this guide to SolidWorks basics include detailed explanations as well as comprehensive visual aids. Each of the projects is presented through clear step-by-step instructions, screen shots, and links to online instructional videos that can be accessed by scanning QR (quick response) codes, or, for ebook readers, by clicking on a direct link. (For a list of these videos, see page vii.)

To help readers develop and expand their SolidWorks skills, the various projects are designed to be followed in sequence. While particular steps can be studied independently, the best way to gain a command of this program is to start at the beginning and work your way through each section in order. With each new section, you will apply the knowledge that you previously learned and build new skills on that foundation.

If you find that you do not recall how to use a command, use the table of contents or the index to find the exercise addressing the specific command and then follow the steps again. Again, you also can watch the online instructional videos, which show how things are done. Plus the appendixes include useful information on contraint/relationships, dimension tolerance types, common short cuts, and mathematical calculations that can be performed in any of the entry fields.

Finally, the contents of this book explain how to use many of feature of SolidWorks, including those found in the 2014 and 2015 versions. In the 2015 release, the majority of commands are unchanged as it impacts this work. However, where there is a difference, this is clearly indicated. The appendix also includes a rundown on a dozen new features of SolidWorks 2015.

For More Information

Adopting instructors will find additional materials for SolidWorks projects at Fred Fulkerson's author's Web site, **sites.google.com/site/fredumacations/solidworks-minions**.

Anyone wishing to learn more after completing this book, may want to work through a free SolidWorks eBook about creating weldments, available at the Industrial Press Web site **ebooks.industrialpress.com/product/design-weldments-using-solidworks-2014**.

SolidWorks
BASICS

A Project Based Approach

1: Getting Started

Introduction

SolidWorks is a full, two-way parametric design tool. Models are fluid in that they may be changed in shape or size, and the linked documents (such as the drawing) will update to reference those changes. Drawing dimensions may also be changed with the subsequent change occurring on the model and associated files. Apart from the dimensional changes that can occur, features may be geometrically linked to others; as a result, one surface will always remain parallel to another and the depth of a slot will always remain exactly one-half of the overall part thickness.

Models are made up of features. These consist of both "part" features that are first sketched and then extruded, revolved, swept, or lofted, and "placed" features (holes, fillets, chamfers, and patterns) that are added to a pre-existing part. Parts are broken down into logical features that are created to form the whole. Part features may either be added to or subtracted from the base body of the part.

Added to this is the ability to find the physical properties of the model or to animate the model showing its complete functioning within the assembly model. Photorealistic images using PhotoWorks can also be produced from the parts or assemblies to help visualize the finished product. Loads can be placed onto the model to determine its resistance to bending or to analyze the weak spots on the part. Individual parts may also be exported to other programs that may aid in applications such as machining or air flow analysis.

Starting SolidWorks

- Select **Start**, **All Programs**
- Scroll to the SolidWorks folder, then select the **SolidWorks** folder.

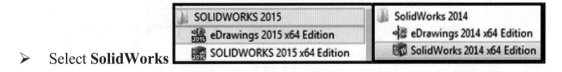

- Select **SolidWorks**

When the software first opens, your screen should look similar to Figure 1.1 on the next page.

2 Chapter 1: Getting Started

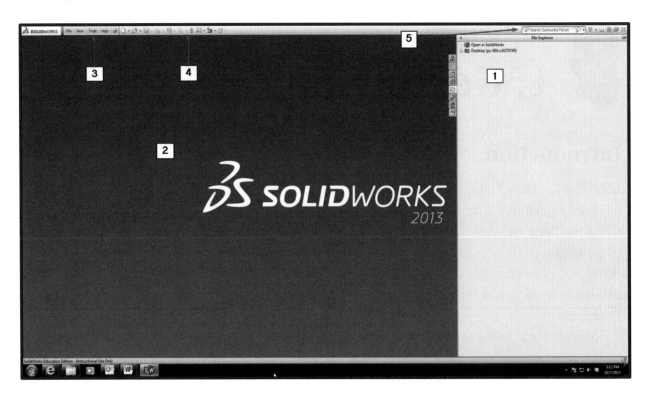

Figure 1.1

User Interface

In Figure 1.1:

- Area 1 is the SolidWorks **"Task Pane"** or file explorer. Here you can:
    - preview your model before opening it
    - select standard items such as fasteners to load into your assemblies
    - access SolidWorks online tutorials
    - access SolidWorks chat groups
    - access a SolidWorks forum
    - start a new document
    - get help
    - watch an instructional video
    - and much more
- Area 2 is the drawing section or graphics area where you will create and manipulate your models.
- Area 3 contains the standard drop down menus, the pin allows you to lock this menu so that it can always be seen, or to unlock it so that it will expand when your mouse is focused over the arrow beside the SolidWorks logo:

Chapter 1: Getting Started 3

- Area 4 contains drop down menus allowing you to perform common commands such as:
    - start a new document
    - save
    - print
    - re-build your part
    - set system options

- Area 5 contains:
    - the online forum search help menu
    - the SolidWorks help menu
    - the standard windows minimize, full screen, and close window options

Areas 3 and 4 make up the **Quick Access** ribbon menu. The commands available from these menus will vary, depending on what you are working on, and can be customized.

Starting a Model

➢ In the **Quick Access** menu select **New**   or select **File, New** or **CTRL + N**. A window similar to that shown in Figure 1.2 will open.

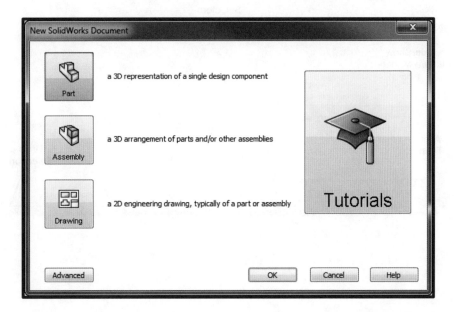

Figure 1.2

4   Chapter 1: Getting Started

From this menu, there are three main options:

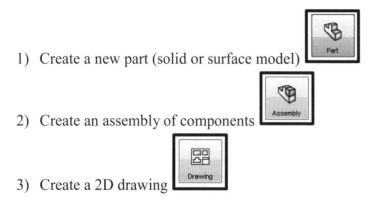

1) Create a new part (solid or surface model)

2) Create an assembly of components

3) Create a 2D drawing

➢ Select **Part**

➢ Select **OK**

Your screen should now look similar to Figure 1.3.

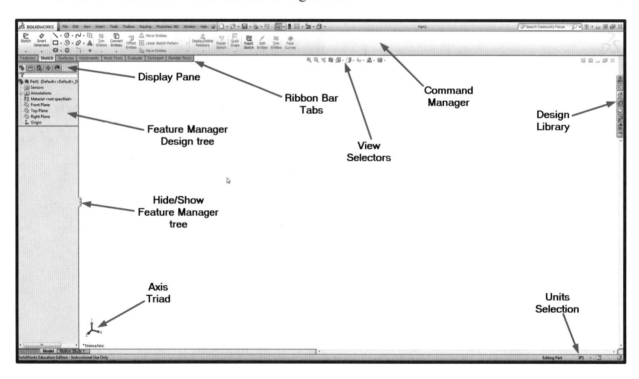

Figure 1.3

Feature Manager Design Tree

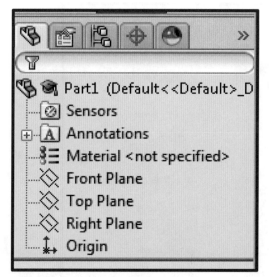

The feature manager design tree shown here in Figure 1.4 details the creation of the part/assembly. It identifies the steps that have taken place to get to where you are and is located on the left side of the screen.

Figure 1.4

Working from the top of the feature manager design tree down, you will notice:

- The **part name**, **Part 1** in this case. The graduation cap [Part1] beside it tells you that this part was created using the educational version of the software. Whatever name you save your part as will display here.
- **Sensors.** This folder will contain any sensors you add to your part or assembly.
- **Annotations.** This folder will contain controls to turn on/off various annotations, such as dimensions.
  Material. By right clicking here, you can select the material for your part.
- **The three standard work planes and the part or assembly origin.** These are default planes on which you can create sketches. They are very useful in that they ensure a part begins on the correct plane and that it is positioned correctly to allow economical creation. *(You should ALWAYS use these planes and the origin axis to position your initial sketches.)*

As you create new features, each will show up in this feature manager design tree.

6 Chapter 1: Getting Started

Reference Planes

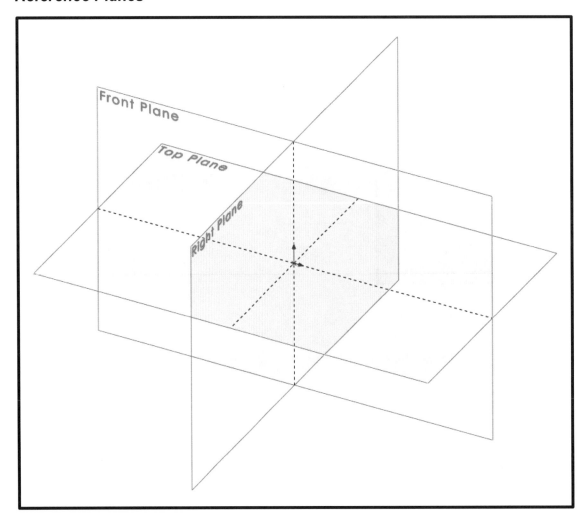

Figure 1.5

The three main reference planes are at 90 degrees to one another. Where they all meet is your part origin. **You must** always lock your parts and assemblies to the part origin. The visibility of any plane can be turned on and off.

> ➢ Right click on the **Front** plane in the feature manager tree.
>
> ➢ Select the glasses , which act as a toggle switch to turn the visibility of the plane on and off. Once the plane is visible as shown in Figure 1.5 above, you can also click on it in the graphics area to select the glasses and hide it.

Turn the visibility on/off of the three planes.

To see this being done, watch the video "**Show Hide Planes SW**."

Tools and Toolbars

Hide/Show Items

In the top middle of the graphics area is the **View (Heads Up)** tool bar and roughly in the middle of this tool bar is **Hide/Show Items** 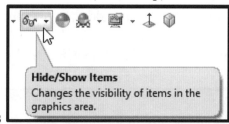 quick selection menu.

From this menu, you can quickly show or hide many commonly used elements:

#1 – planes

#2 – origins

#3 – axes

#4 – temporary axes

#5 – coordinate systems

#6 – points

#7 – curves

#8 – parting lines

#9 – 3D sketch dimensions

#10 – all annotations

#11 – sketches

#12 – grid

#13 – sketch relations

#14 – 3D sketch planes

#15 – lights

#16 - cameras

#17 – decals

#18 – live section planes

#19 – routing points

#20 – dimension names

#21 – view simulation symbols

#22 – weld beads

#23 – center of mass

8 Chapter 1: Getting Started

The Sketch Ribbon Bar

All parts will start with a sketch. It is very important that you understand SolidWorks is a true sketching tool. This means that when you start sketching, you should not worry about exact sizes or shapes. All sketches should be made as if you were sketching on a pad of paper. The dimensions and specific constraints will be added later to fully define your parts.

The sketch panel, shown in Figure 1.6, appears when you are in sketch mode.

Figure 1.6

Note how each of the tabs is a context sensitive menu. Grayed out selections are not available to be used at that specific time.

- ➢ Once you are comfortable with the sketch commands (or any ribbon bar commands), you may wish to use the **expert** option, where the text descriptions of the commands are removed as shown in Figure 1.7.
    - o Right mouse click on the sketch panel,
    - o Select **Use Large Buttons with Text** (uncheck this option). The sketch ribbon should now look like Figure 1.7.

Figure 1.7

For now, let's turn the large buttons with text back on.

- ➢ Right mouse click on the sketch panel.
- ➢ Select **Use Large Buttons with Text**

Docking the Menu Bar

- Click and hold on the **Sketch** menu tab
- Drag it out into the middle of your screen.
- While still holding the mouse, you should notice three arrows: one near the top center of your screen, one near the center right, and one near the center left of your screen, as shown in Figure 1.8.

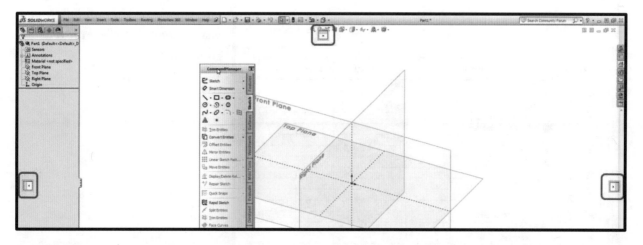

Figure 1.8

- Position your mouse over one of these arrows and release it. The **Command Manager** is now docked. You also can leave the Command Manager free in the graphics area (try it in several locations until you decide what you prefer).

To see this being done, watch the video "<u>Docking the Menu Bar</u>."

Double-clicking on the floating menu bar will also re-dock the Command Manger automatically.

Shortcut Keys

- Right click in the menu. A new window will open as shown in Figure 1.9.

Figure 1.9

10 Chapter 1: Getting Started

> Select **Customize**

> Select the **Keyboard** tab

You will now see a window similar to Figure 1.10.

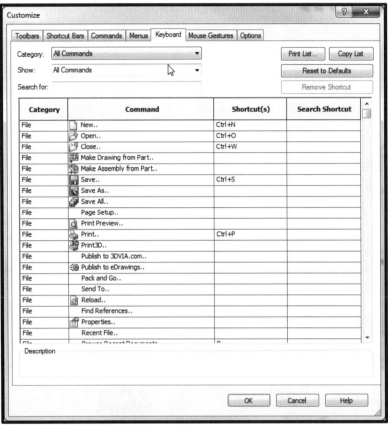

Assign shortcut keys as you see fit, but be certain to first confirm that your new key combination has not been used for another command.

Figure 1.10

> Accept these selections when complete

Pressing the **S** key will show the shortcut menu at the cursor location on your screen. Right mouse button click in this shortcut menu to allow you to add icons for often used commands.

Pressing the **R** key will bring up the recent documents, allowing you to open recently closed documents.

The following **Windows** short cut keys may also be used:

Keys	Result
Ctrl+C	Copy selected items.
Ctrl+N	Create a new document.
Ctrl+O	Open a new document.
Ctrl+P	Print the active document.
Ctrl+S	Save the active document.
Ctrl+V	Paste items from the clipboard to the active document.
Ctrl+Y	Redo.
Ctrl+Z	Undo.

Adding Commands to a Ribbon Bar

➢ Right click in the menu, and a new window will open as shown in Figure 1.9.

➢ Select **Customize**

➢ Select the **Commands** tab

A new window will open that look similar to Figure 1.11.

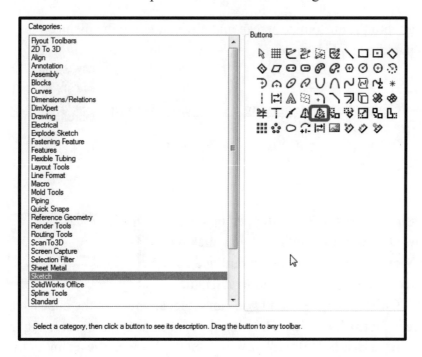

Figure 1.11

*The left side of this window lists all of the **Categories** of commands; from this list, you will select the desired category. The right side shows all of the **Buttons** (icons) available for this category. To add a button to a toolbar, simply click on the icon and drag and drop it into the desired toolbar.*

➢ Select the **Sketch** category as shown in Figure 1.11 above.

12 Chapter 1: Getting Started

- Select the **Dynamic Mirror Entities** icon, as shown in Figure 1.11 on the previous page.
- Drag and drop it into your **Sketch** toolbar so that it looks similar to Figure 1.12.

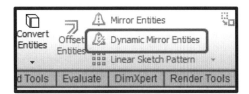

Figure 1.12

- Accept this selection by hitting

To see this being done, watch the video "**Adding a Command Button**."

To remove a menu icon, go into the customize screen and click and drag the menu icon from the menu bar a drop it into the graphics area.

To see this being done, watch the video "**Removing a Command Button**."

View Orientation

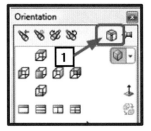

Pressing the **Space Bar** will open up the **View Orientation** window. If you click on the **View Cube** option (shown outlined by the red box with the #1 pointing at it), the

View Cube will appear, and you can select any face of the cube as the current "normal to" view. Pressing the space bar again or selecting the **View Orientation** from the **View (Heads-Up)** toolbar will open the **View Cube** again, until you click on the **View Cube** option again to shut it off. See Figure 1.13 on the next page.

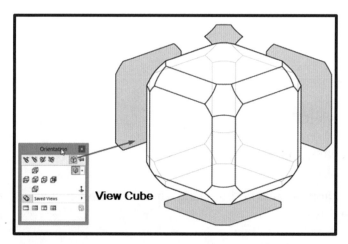

Figure 1.13

Mouse Gestures

Holding the RMB and moving the mouse will bring up common commands directly related to the current active command state. For example, if you have a blank part with nothing started, the  view orientation will populate the mouse gesture options.

If you are in an active sketch, the common sketching commands will populate the mouse gestures. To select the desired option, simply move the mouse over that section.

The mouse gestures can be customized by:

➢ Right click in the menu, and a new window will open as shown in Figure 1.9.

➢ Select **Customize** Customize...

➢ Select **Mouse Gestures** Mouse Gestures

Make any changes that you want to customize the mouse gestures, and then save your changes.

File Types

There are seven basic file types:

File Extension	File Type
.sldprt	part file (to create individual models)
.slddrw	drawing file (to create drawing files from models)
.sldasm	assembly file (a collection of parts and/or standard components mated together into an assembly)
.prtdot	part modeling template file (a part template created to always include common elements, such as your name, company, and so on, into your part properties)
.drwdot	drawing template file (a drawing template created to load common elements into your drawing sheets)
.asmdot	assembly modeling template file (an assembly template created to load common elements into your assemblies)
.slddrt	SolidWorks sheet format

Backing up Files

Your files should be backed up regularly to prevent losing valuable data. The default back up location may not be convenient for your needs if you are working on a network computer with "**Deep Freeze**." If this is the case, follow these steps to change the location of where your files are backed up. *(If you do not have the **Deep Freeze** software, back up your files otherwise.)*

- Select **Tools**
- Select **Options**
- In the **System Options** general window, select the **System Options** tab
- From the column on the left-hand side of the window, select **Backup/Recover**

➤ In the **Backup** [Backup] section, activate the radio button to **Save backup files in the same location as the original** option [⦿ Save backup files in the same location as the original]

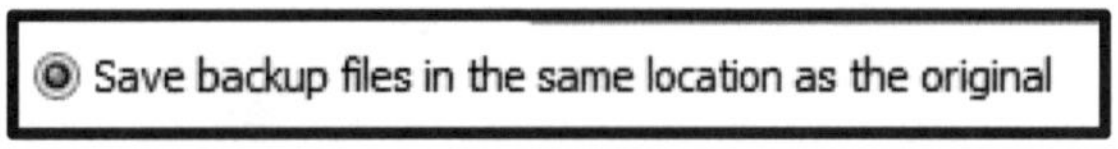

➤ Set up the **Auto-recover** [Auto-recover] 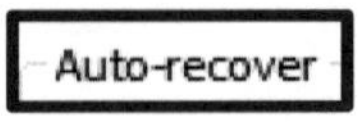 section as shown in Figure 1.14 for a computer with "**Deep Freeze**." *(For any other computer, regularly save your files to the location you choose.)*

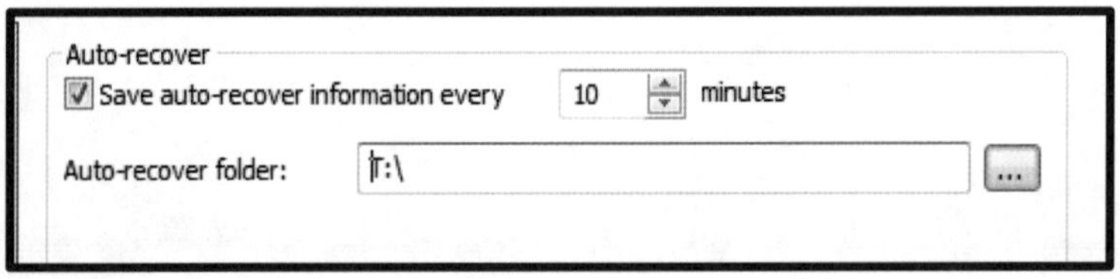

Figure 1.14

➤ Select **OK** [OK]

2: Basic Sketching

Line/Arc Sketching

Complex shapes, which include both lines and arcs, can be created by using just the line sketching tool. The handle shown in Figure 2.1 will be created by this method.

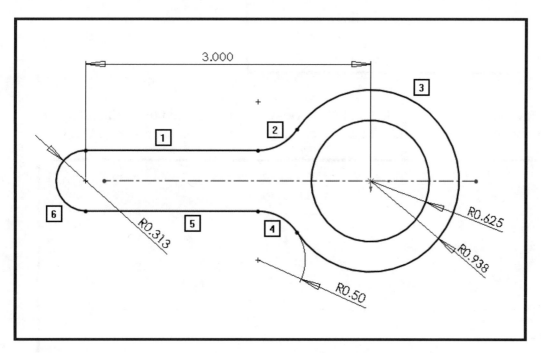

Figure 2.1

- Start a new SolidWorks inch part

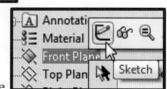

- Start a new sketch in the front plane
  - Right click on the **Front Plane** in the **Feature Tree.**
  - Select **Sketch**

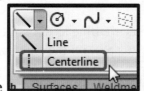

- Select **Centerline**

17

18 Chapter 2: Basic Sketching

➢ Create a center line similar to that shown in Figure 2.1.
  o Position your mouse in line with the origin, as shown in Figure 2.2. Click to start the center line.

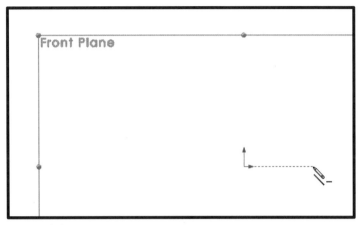

You should see the dotted line indicating that your new line will be in line with or coincident to the origin and the horizontal line constraint at the end of the pencil.

Figure 2.2

  o Move your mouse to the left and click. Be careful to maintain the **Coincident** and horizontal line constraints. Your screen should now look similar to Figure 2.3.

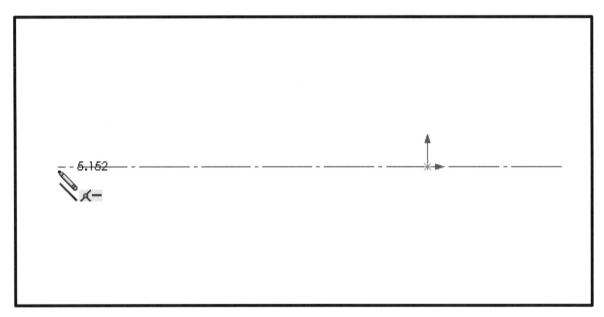

Figure 2.3

  o Press the **Esc** key to exit the command.

You should now notice two important items:
1) Your line is black in color but the end points are blue as shown in Figure 2.5 on page 20.
2) Across the bottom of your screen in the status bar, on the right-hand side, you will see that your sketch is **Fully Defined** [Fully Defined]

All of your sketches, at this point, should be fully defined. *(If your sketch is not fully defined, then drag any blue entity end point around to see if a relationship and/or a dimension is missing.)* When an entity turns black, it is fully defined; if it is blue, then it is under defined (meaning that it can still be changed); and if it is red, it is over-defined (meaning that two or more dimensions and/or relationships are generating conflicting solutions).

So why does the system tell you that this line is fully defined if the end points are blue? Since this line is a center line, the length does not need to be defined. *(Because it will be used as an axis of symmetry for your sketch, its length is irrelevant.)*

➢ Click on one of your end points and hold down the mouse button. Now, drag the end point around to change its length.

➢ Click (select) on the line. Your screen should look similar to Figure 2.4.

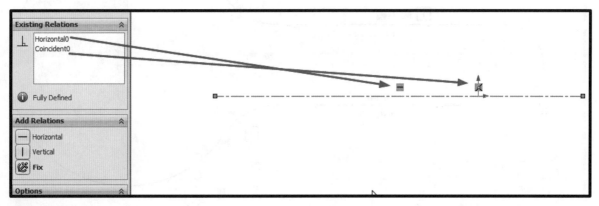

Figure 2.4

This line currently has two **constraints** that fully define it: the **horizontal** (meaning the line is horizontal) and the **coincident** (meaning the entity is in line with another object, in this case, the origin)

➢ Activate the **Line** tool, either by clicking on the **Line** icon or by selecting **Tools, Sketch Entities, Line:**

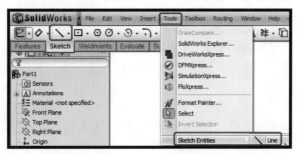

➤ Sketch line 1 as shown in Figure 2.5. *Remember that the exact location and length is not critical. When sketching, also remember that you need to click your mouse in the graphics area to start the sketch entity and to end the sketch entity.*

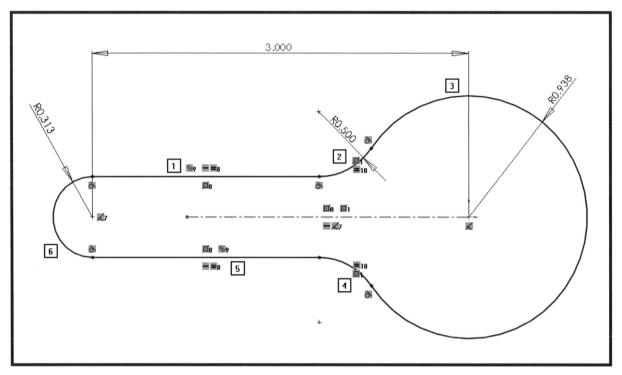

Figure 2.5

➤ Press the **A** key, SolidWorks will now switch to the three-point arc command. Sketch arc 2. **Do not worry about size, just work on getting the approximate shape.**

If the arc is not being created as you think it should be, then move your mouse over the end point. This will change how the arc is created.

➤ Press the **A** key again and sketch arc 3.
➤ Continue sketching this part until it is complete, as shown in Figure 2.5.

If you are not sure how to do this, watch the video "<u>LA Handle</u>."

Fully Defining the Handle

When fully defining a part by adding dimensions and sketch relations, it does not matter if the dimensions or the relations are added first. There are also many different ways to constrain this part other than the ones described below.

> Add the dimensions to your sketch as shown in Figure 2.5 on the previous page.

- Select **Smart Dimension**
- Select one of the arcs. Then move the mouse and click to place the dimension.
- Immediately enter the desired feature size in the **Modify** dimension window 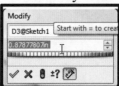 and accept your input by pressing the **Enter** key or clicking on the check mark
- Select the R0.938. Then select the R0.313 arc to add the 3.0 center to center dimension.

*When dimensioning features, you should notice that when the **Modify** dimension window opens up, the input field is highlighted blue. Therefore, all you need to do is to input the desired feature size and accept your input by pressing the **Enter** key. You can also enter an equation in this field by first typing the equals sign and then the equation to define the feature size.*

> Constrain the center of the R0.938 circle to the sketch origin.
- Select the center of the circle (click on it).

- Hold the **Ctrl** (or the **Shift**) key and select the origin

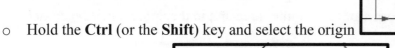

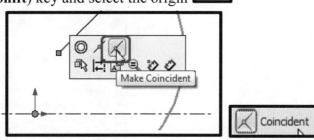

- Select the **Coincident** constraint.

*After you select the second entity, if you do not move your mouse, the possible constraints will pop up and the **Coincident** constraint can be selected from this pop up window. If you do move your mouse, select the constraint from the **Add Relations Properties** on the left side of your screen.*

22 Chapter 2: Basic Sketching

➢ Add the **Coincident** relationship between the center line and the center of the R0.313 circle.

Notice how the features are turning black as they become fully defined.

➢ Add the **Equal** 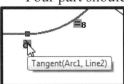 relationship to arcs 2 and 4 as shown in Figure 2.5 above.
➢ Add the **Equal** relationship to lines 1 and 5.

Your part should now be fully defined. If it is not, you may need to add the **Tangent** 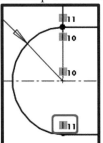 relationship between the lines and the arcs or between the arcs. Or you may need to add the **Vertical** relationship to the end points and center point of the R0.313. Add the necessary constraints until your part is fully defined.

To see this being done, watch the video "**LA Constraints**."

➢ **Exit** the sketch by selecting the **Exit Sketch** command. It can be found near the upper right-hand corner of the graphics area and in the upper left-hand corner.

In the **Feature Manager Design Tree** , you will see your newly created sketch at the bottom of the list.

There are a number of things that you should be aware of at this point:

1) If there is a **Negative sign** [(-) Sketch1] in front of the sketch name, then the sketch is not fully defined. This is a dangerous way to leave a sketch, as you can easily alter a non-fully defined sketch without realizing what you have done.

2) If you right click on the feature **Sketch1**, many options will appear (right clicking will always bring up helpful menus).

3) After right clicking on the feature **Sketch1**, you can choose to edit the sketch by selecting **Edit Sketch** [Edit Sketch]

4) If you started the sketch in the wrong construction plane, you can change the sketch plane by selecting **Edit Sketch Plane** [Edit Sketch Plane]

5) You can also scroll down and select **Feature Properties** [Feature Properties...] to see what user was logged [Created by: ffulkerson] onto the computer and the exact date and what time the feature was created [Date created: 10/24/2013 8:40:27 AM / Last modified: 10/24/2013 8:40:27 AM]

6) If you perform a slow double click on the feature name, you can rename it.

7) The symbol beside the sketch in the feature tree will vary, depending on several factors. If the sketch is shared by more than one feature, it will show a hand holding it. If it has more than one enclosed shape, this symbol will change to a polygon with a hole in it. Currently, this symbol tells you that it is a single fully defined shape. If there is a negative sign on the left side, then the sketch is not fully defined.

24 Chapter 2: Basic Sketching

Renaming the Sketch

- Right click on **Sketch1** in the feature tree.
- Select **Feature Properties**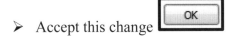
- In the **Name** field enter **Handle**
- Accept this change

Your feature tree should now look similar to Figure 2.6.

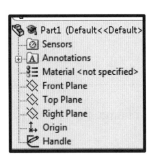

Figure 2.6

Extruding the Handle

- Select the **Features** tab
- Select the sketch by clicking once on it in the **Feature Manager design tree** (#1) as shown in Figure 2.7.

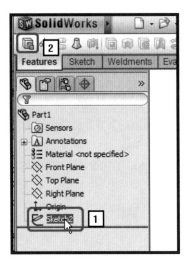

Figure 2.7

Chapter 2: Basic Sketching **25**

- Select **Extruded Boss/Base** . The **Boss-Extrude** window will now open, (#2) as shown in Figure 2.8.

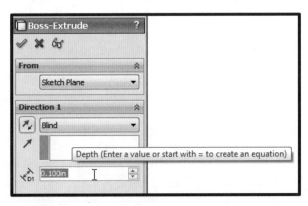

Figure 2.8

You should notice that the **Depth** field is highlighted so that you can immediately enter the thickness for your part.

- Enter **0.5** as the **Depth.** You should notice the preview of your part change once you press the **Enter** key.

- Accept this selection

To see this being done, watch the video "**LA Handle Extrude**."

Your part should now look similar to Figure 2.9.

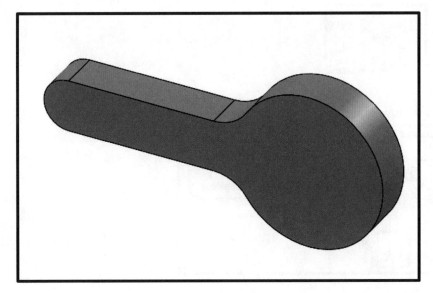

Figure 2.9

- **Save your work.**

Editing your Work

Renaming a Feature

> Click on the **Boss-Extrude1** feature, in the feature tree, wait a second, and click on it again.

Your **Boss-Extrude** feature should now look similar to Figure 2.10. Notice how the name is highlighted in blue and appears in a box.

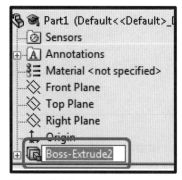

Figure 2.10

> Type **HandleExtrude** (with no space) as the new name

> Click anywhere in the graphics area.

Editing a Sketch

> Right click on the **Handle Extrude** feature (in the feature tree or in the graphics area).

> Select **Edit Sketch**
> Press the **Space Bar**.
> Select the **Normal To** view option:

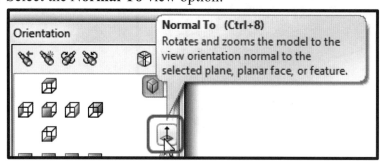

Chapter 2: Basic Sketching 27

- Select the **Circle** command
- Select the **Origin** as the center of the circle
- Move your mouse and click to create a new circle.
- Right click and choose **Select** to exit the command. (This is the same as pressing the **Esc** key.)
- Select **Smart Dimension**
- Select this new circle.
- Move your mouse and click to place the dimension.
- In the **Modify** dimension dialogue box, enter: **=0.625*2**
- Accept this equation.

Your screen should now look similar to Figure 2.11. You should notice the red equation symbol beside the dimension and a new folder titled **Equations** in the feature tree.

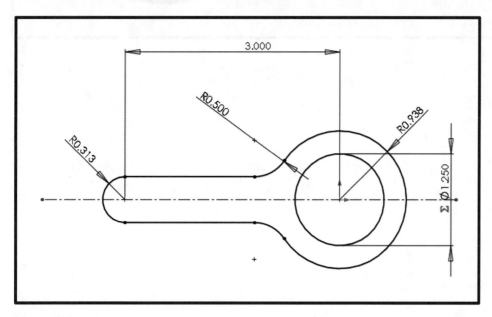

Figure 2.11

After you make your changes, you should notice as asterisk beside the part name Block.SLDPRT* that lets you know that your changes have not been saved.

- **Save your work.**

28 Chapter 2: Basic Sketching

➢ Exit the sketch.

Your part should now look similar to Figure 2.12.

Figure 2.12

3. Custom Templates

The environment that we work within while creating models, drawings and presentations can be controlled. For instance, you may want automatic retrieval of dimensions to take place, have specific creation icons show on existing or user-defined menus, or control the background colors of the screen. The drawing file may also require some explicit title blocks, dimension styles, or logos to appear.

All of these are controlled by **Template** Files. These are pre-existing start-up files that contain the set options you need—they are not overwritten by the part or drawing creation but can be saved as template files so that customization of your working environment can take place. You can save yourself a considerable amount of time if you take a few minutes to customize SolidWorks. **The following steps will ONLY be done ONCE**.

➤ Download the file **$SolidWorksSetup.zip**.
➤ Extract the files to your computer. *(If you are working on a network drive, then extract the folder to your network drive space.)*
➤ Position your mouse over the arrow beside the word **SolidWorks** in the upper left-hand corner of your screen and expand the menu

Now, we need to tell **SolidWorks** where to find these templates.

➤ Select Tools, Options
➤ Select **File Locations** from the **System Options** tab as shown in Figure 3.1.

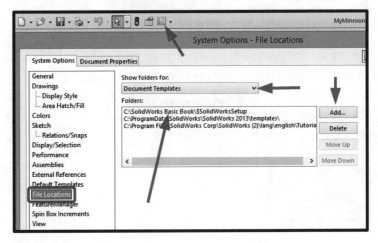

Figure 3.1

29

30 Chapter 3: Custom Templates

- Select the **Document Templates** location from the drop down list. (It should be the default.)
- Select **Add** and navigate to the location to your **$SolidWorksSetup** folder.

From this point forward, every time you start a new part, you will use your templates by selecting the **Advanced** tab and then the **$SolidWorksSetup** tab in the **New SolidWorks Document** window as shown in Figure 3.2.

Figure 3.2

Customizing Your Templates

- Open **SolidWorks**
- Open the **Task Pane** on the right-hand side of your screen. This can be done by clicking on the **File Explorer** tab
- Expand the desktop by clicking on the plus sign beside the word **Desktop**
- Navigate to where you saved the **$SolidWorksSetup** folder and expand it.
- Click and hold on the **ANSI-MM.prtdot** file
- Drag and drop it the graphics area.
- Click in the graphics area.
- Select **Files, Properties**

- In the **Summary Information** window, select the **Summary** tab. Add **Your Name** as the 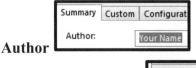 **Author**
- Select the **Custom** tab
- In the tenth row of the **Value/Text Expression** column, enter **Your Name**

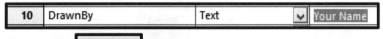

- Select **OK**
- **Save your changes.**
- Close the template
- Select **Save all.**

Now, every time you start a new metric part, your name will be included with the part. ***After you edit each template, never open them this way again.*** *You should also never save anything else to your $SolidWorksSetup folder.*

- Follow these steps again to add your name to the **ANSI-MM.asmdot** (the metric assembly template), **ANSI-INCH.prtdot** (the inch part template) and the **ANSI-INCH.asmdot** templates.

Drawing Templates

- Drag and drop the **ANSI-A-Size-MM.DRWDOT** template in the graphics area.
- Right click in the graphics area and select **Edit Sheet Format**
- Double click on the writing in the **PROPRIETARY** section of the title block and change the statement to reflect your company name **YOUR COMPANY.**
- Click in the white space of the drawing sheet template.
- Double click on the company information section (the upper right-hand cell of the title block) and replace **THE MIDDLE OF NOWHERE** with your address.
- Select the picture of the pink ribbon and press the **Delete** key on your keyboard.
- Confirm the deletion of the picture.

- Select **Insert, Picture**
- Browse to a location on your computer where you have saved your company logo.
- Select the image file and select **Open**.
- Select the **fit screen** option
- Resize your picture by dragging one of the corners
- Then move the picture, by clicking on it and dragging it, to where the ribbon was.
- Select **OK** 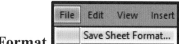 to close the picture and **stop editing it**.
- Stop editing the sheet by right clicking in the sheet and selecting **Edit Sheet** or by selecting the icon in the upper right-hand corner of the graphics area.
- **Save your changes.**
- Select **File, Save Sheet Format**
- Navigate to your **$SolidWorksSetup** folder.
- Save the file as **ANSI-A-SIZE-MM.slddrt**
- Close the template.
- Select **Save all**.
- Repeat these steps for the remaining **DRWDOT** inch and metric templates, sheet sizes A through to E.

Do not edit any of the blue writing in the templates. These are the links of the properties from the part and assembly templates where you added your name.

Toolbox Customization

If you are working on a network and using different computers in different rooms, you will also want to configure your toolbox. The toolbox contains numerous standard components—bolts, nuts, washers, bearings, and so on.

- ➤ Select the drop down arrow beside **Options.**
- ➤ Select **Add-Ins**
- ➤ In the **Add-Ins** window, check in the boxes on each side of the **SolidWorks Toolbox** and **SolidWorks Toolbox Browser**
- ➤ Accept these selections [OK]

Note: The more add-ins you turn on, the longer it will take SolidWorks to start up. So do not turn on an add-in unless you are going to use it.

- ➤ Where you created your **$SolidWorksSetup** folder, create another empty folder called **$ToolboxParts**.
- ➤ Start a new part.
- ➤ *For SolidWorks 2014:* from the drop down menu across the top, select **Toolbox Configure**

For SolidWorks 2015: from the drop down menu across the top, select **Tools** [Tools], **Toolbox** [Toolbox], **Configure** [Configure...]

- ➤ When the **Toolbox** configuration window opens, select step 3, **Define user settings**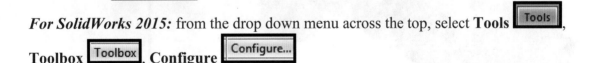

- ➤ In the **Files** sections, select the radio button for **Create Parts**: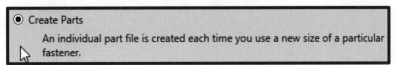

34 Chapter 3: Custom Templates

- Select the icon with the three dots in the **Create parts in this folder** box:

- Navigate to where you created your **$ToolboxParts** folder and select it.
- **Save your changes.**
- Close this window.

If you do not bother to setup the toolbox this way, then when you open an assembly, the toolbox will try to look on the C drive of the computer where the assembly was first created. If you are not on the same computer, it will not be able to find the correct part. This will cause your assembly to look similar to Figure 3.3.

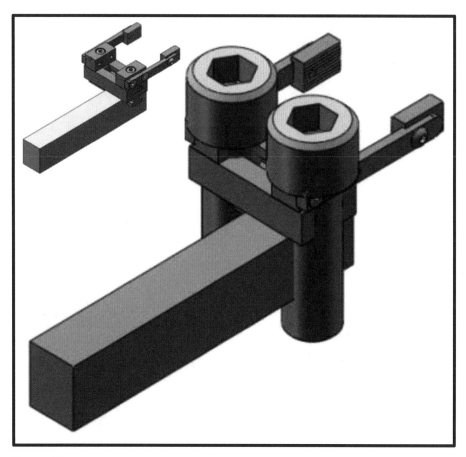

Figure 3.3

Copy Settings Wizard

Saving Your Settings

If your school or workplace computers have **Deep Freeze** software installed, you will need to save a copy of all of your settings in your network drive space and reload it **each time** you start using SolidWorks. As long as you do not have Deep Freeze at home, you will not need these extra steps, but if you want to copy your setting to a different computer, then this is one way.

- *For SolidWorks 2014:* close all open documents, but **do not exit SolidWorks.**
- *For SolidWorks 2015:* you do not need to close the document(s).

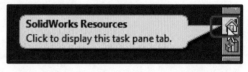

- Select the **SolidWorks Resources** tab found on the right side of your screen.
- In the **SolidWorks Tools** section, select the **Copy Settings Wizard**

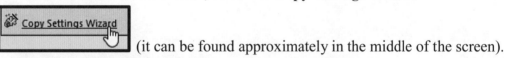

(it can be found approximately in the middle of the screen).

- Select **Save Settings,** then **Next:**

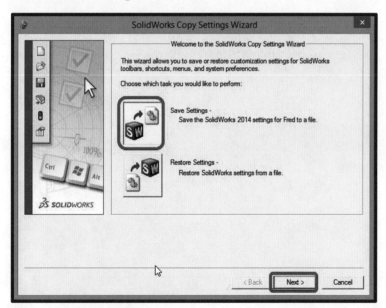

- Browse [Browse] to your **$SolidWorksSetup** folder and name the file **swSettingsYourName.sldreg**.
- Select **Finish** [Finish] and then select **OK**.

Restoring Your Settings

If you want to copy your custom settings to a memory stick and install them on a new computer or copy them each time you first use SolidWorks on a computer with **Deep Freeze**, you will need to run the copy settings wizard. This time, just use the **Restore Settings** option and select your settings file.

 4: LEGO Minions, Part 1—Designing

2 x 2 Block

> Start a new SolidWorks part and set your units to millimeters:

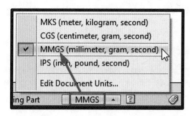

Square Base Extrusion

> From the **Features** ribbon bar, select **Extruded Boss/Base**

> Select the **Top Plane** as the sketch plane

SolidWorks should now have rotated the top plane "normal to" the computer screen and have made the **Sketch** tab active

Rectangle

> Select the down arrow beside the Rectangle
> Select the Corner Rectangle option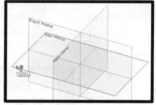
> Select the origin as the first corner.

37

- Move the mouse in quadrant 4, lower and to the right, and click for second corner location. Your screen should look similar to Figure 4.1.

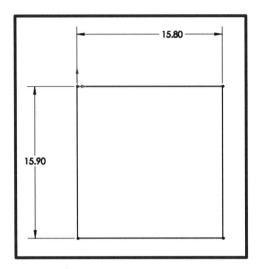

Figure 4.1

Dimension

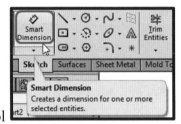

- Select the **Smart Dimension** tool
- Select one of the horizontal lines, move your mouse above the rectangle, and click to place the dimension.

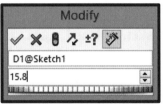

- Enter **15.8** as the feature size and accept this selection
- Select one of the vertical lines and make it **15.9** in length.
- Your sketch should now be **Fully Defined**
- Exit the sketch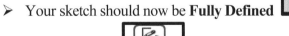
- **Save your work.**

The extrusion window should now be open on the left-hand side of your screen and look similar to Figure 4.2, unless you have exited the sketch and re-opened it. If the extrusion window is not open, select your sketch, and then select extrusion.

Figure 4.2

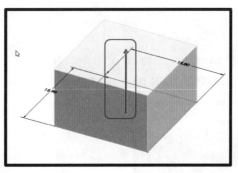

- Make the extrusion direction up

- Set the distance to 9.6

- Turn the draft on and set the angle to 1 degree

40 Chapter 4: LEGO Minions, Part 1—Designing

➢ Accept these selections ☑. Your screen should now look similar to Figure 4.3.

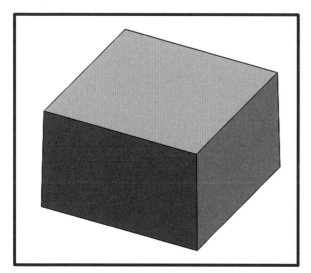

Figure 4.3

➢ Save your part as **TallBlock_FirstName_LastName**.

Round Extrusions

➢ If your **Features** tab is not active, select it.
➢ Select the top face of your part as shown in Figure 4.4.

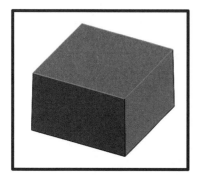

Figure 4.4

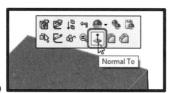

➢ Select **Normal To**
➢ Select **Extruded Boss/Base.**

You are now creating a sketch on the top of the part for the next extrusion.

Chapter 4: LEGO Minions, Part 1—Designing **41**

➢ Select **Circle** as the sketching tool.
➢ Click near the lower left-hand corner of the piece to set the location of the circle's center.
➢ Move the mouse away and click again to create a circle.
➢ Press the **ESC** key to exit the circle command.
➢ Use the **Smart Dimension** tool and dimension the circle at **4.8 mm** in diameter by clicking anywhere on the circle.
➢ Dimension from the left side of the part to the center of the circle, **3.9 mm**, by clicking on the farthest part edge and then clicking anywhere on the circle.
➢ Dimension **3.9 mm** from the bottom of the part to the circle center.
➢ Press the **ESC** key to exit the command.

The sketch should now be fully defined.

➢ Exit the sketch

➢ Set up the extrusion window as shown in Figure 4.5. **DO NOT** uncheck the **Merge result,** as you want this to stay as one solid part.

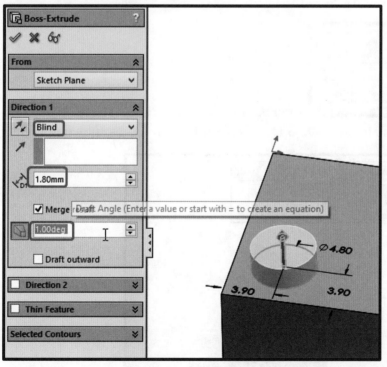

When dimensioning, be careful to select the bottom edge of the part and not the top edge. Due to the 1 degree draft, these will be close to one another.

Figure 4.5

➢ Accept these selections

42 Chapter 4: Lego Minions, Part 1—Designing

➢ Your screen, in the isometric view, should now look similar to Figure 4.6.

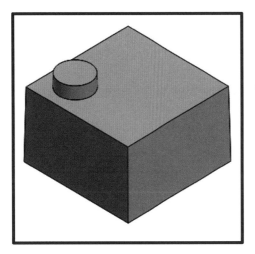

Figure 4.6

Feature Fillet

If your **Features** tab is not active, select it.

➢ Select **Fillet** from the **Feature** menu.
➢ Set up the fillet menu to create **0.25 mm Constant size** fillets as shown in Figure 4.7.

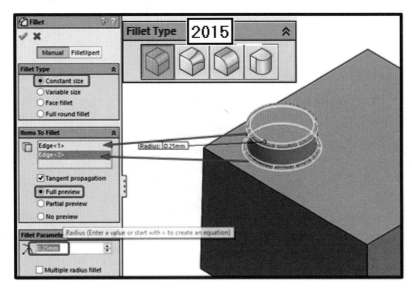

Figure 4.7

- Select the edges of the part as shown.
- Accept these selections ✓
- Add another **0.25 mm** fillet feature. Your part should now look similar to Figure 4.8

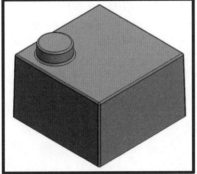

Double click on your part in the graphics area, and you should notice that all of your dimensions will appear. The black dimensions come from the sketch, and the blue dimensions come from the features.

Any of these can be changed, without editing the sketch or feature, by double clicking on them and then rebuilding the part.

Do not fillet the bottom edges of the part.

Figure 4.8

Linear Pattern

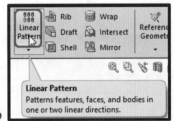

- Select **Linear Pattern** from the **Features** tab
- Select a horizontal (parallel to the X axis) edge of the part, as indicated at #1 in Figure 4.9. This will be the **Direction 1** for the pattern.

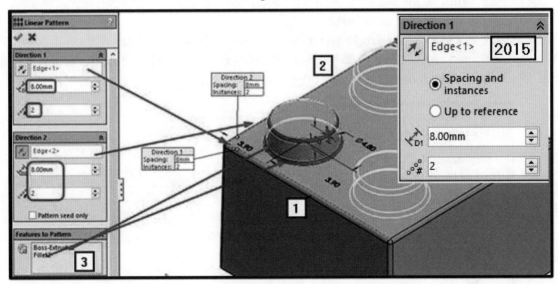

Figure 4.9

44 Chapter 4: LEGO Minions, Part 1—Designing

When selecting the edge, be sure to select close to the lower left-hand corner so that the direction arrow is pointing in the direction that the features will be patterned, toward the lower right-hand corner of the part, in this case.

The direction of the pattern can be reversed either by clicking on the direction arrow in the graphics screen or by clicking on the reverse direction icon as shown in Figure 4.10.

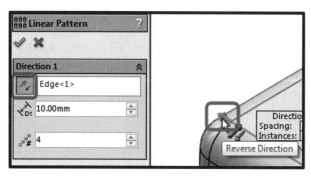

Figure 4.10

> Enter **8 mm** as the distance between each instance.
> Enter **2** as the number of instances.
> Click in the **Pattern Direction 2** cell. Then select a vertical line (parallel to the Z axis), as shown at the #2 in Figure 4.9 on the previous page, to set the second direction for the pattern. (*If the pattern is in one direction only, then leave the second direction field blank.*)
> Enter **8 mm** as the distance between each instance.
> Enter **2** as the number of instances.
> Click in the **Features to Pattern** cell (as shown by #3 in Figure 4.9 on the previous page), and then select the round boss extrude and the fillet feature at the top and bottom of the round extrusion as the features to pattern:

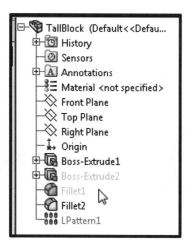

Features can be selected by clicking on them on the part in the graphics area or by selecting them from the feature drop down menu in the upper right-hand corner of the graphics area.

Accept these selections. Your part should look similar to Figure 4.11.

Figure 4.11

➢ **Save your work.**

To see this being done, watch the video "**LPattern**."

Next, we want to shell our part. This command removes material from the selected faces until the remaining material is the desired thickness, much like a snail's shell. Unfortunately, if we shell our part as is, the outcome will not be the desired shape, but will look like Figure 4.12.

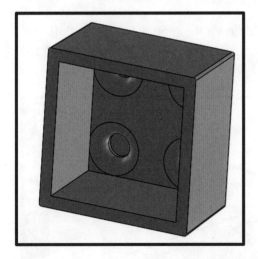

Figure 4.12

46 Chapter 4: LEGO Minions, Part 1—Designing

With the fillets around the bottom of the round extrusion features, SolidWorks shell command will maintain the desired thickness when possible throughout the part, thus creating the large fillets as shown in Figure 4.12 on the previous page. If left as shown, the cost of the mold tool would increase significantly to add these fillets, which serve no purpose. Therefore, before the part is shelled, you will need to rearrange some of the features to avoid this unecessary cost.

Edit a Feature

> In the feature tree, select the **LPattern** feature. It does not matter if you right or left click.

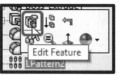

> From the pop up, select **Edit Feature**

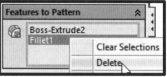

> In the **Features to Pattern** section, select the **Fillet** feature and then press the **Delete** key on your keyboard.

*Alternatively, you can right click on the desired feature name and then, select **Delete** from the options.*

> Accept these changes

Re-Ordering Items in the Feature Tree

> Click on the first Fillet feature that you created, hold down the mouse button, and drag it to the bottom of the feature tree. Then release the mouse button as shown in Figure 4.13.

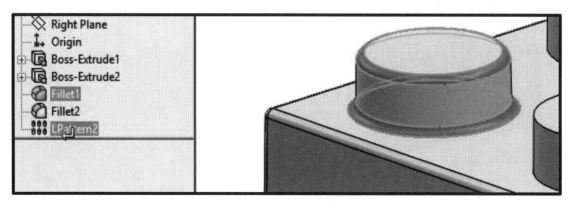

Figure 4.13

You should see the selected feature(s) highlighted in blue on your part. You also should notice the yellow return arrow showing where the feature will be moved to in the feature tree. *If the move is not possible, the yellow return arrow will change to a black circle with a line through it* . *This tells you that there is a parent/child relationship that will not allow the feature to be moved.*

To see this being done, watch the video "**Re_Arranging Features**."

Rollback Bar

> Move the mouse until it is over the **Rollback Bar** (the line normally found below the last feature created) and has the appearance of a hand with the thumb and a finger pinching the line .
> Click and hold the mouse.
> Drag the **Rollback Bar** so that it is above the **Fillet** and below the **LPattern** feature

 and release the mouse button. This will suppress all features below the bar and will allow you to create new features that are not dependant on any feature below it.

To see this being done, watch the video "**Re_Arranging Features**."

Shell

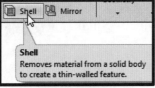

> Select the **Shell** command from the **Features** ribbon bar.
> Set the shell thickness to **1.2 mm**

➢ Select the bottom face of the part as the **Faces to Remove:**

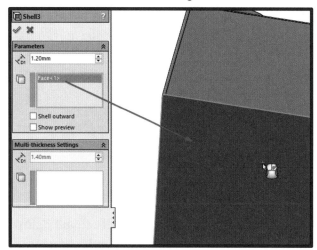

➢ Accept these selections. Your part should look similar to Figure 4.14.

Figure 4.14

To see this being done, watch the video "Shell."

Chapter 4: LEGO Minions, Part 1—Designing   **49**

➢ Move the **Rollback Bar** back to the bottom of the feature tree.

➢ Edit  the **Fillet** so that your part now looks similar to Figure 4.15.

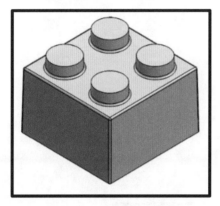

Notice how there is no longer a fillet at the top of the feature.

Figure 4.15

Chamfer

➢ While holding the **CTRL** key, select the top edge of each of the four round features as shown in Figure 4.16.

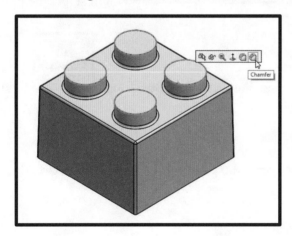

Figure 4.16

You can select the **Chamfer** command from the options as shown, or:

➢ From the **Features** ribbon bar, select **Chamfer**. *(You will first need to select the down arrow found just below the **Fillet** feature option.)*

➢ Set the chamfer type **Angle distance**

➢ Set the size at **0.25 mm** 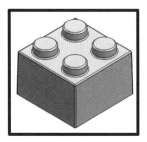, the angle at **45 degrees**, and accept these selections. Your part should now look similar to Figure 4.17.

Figure 4.17

➢ **Save your work.**

Interior Features

➢ Select the bottom face of your part 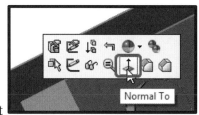 and select the option to look **Normal To.**

➢ With the bottom face still highlighted, select the **Extruded Boss/Base** feature.
➢ Create two circles that share a center point.

Constraining to a Midpoint

➢ Position your mouse over the vertical edge of the part and right click. The part edge will turn orange

In SolidWorks 2015: the edge will turn blue

➢ Choose **Select Midpoint**
➢ Hold the **CTRL** key and select the center of the circle.
➢ Select the **Horizontal** constraint.

To see this being done, watch the video "**Midpoint Constraint**."

- ➢ Position your mouse over the horizontal edge of the part and right click. The part edge will turn orange

  In SolidWorks 2015: the edge will turn blue

- ➢ Choose **Select Midpoint**
- ➢ Hold the **CTRL** key and select the center of the circles.
- ➢ Select the **Vertical** relationship. You should notice that the center of the circles has now changed to black indicating that its position is now fully defined. However, the circles will remain blue until they are dimensioned or further constrained.
- ➢ **Dimension** these circles at **4.8** and **6.514 mm**. The sketch should now be **Fully Defined**:
- ➢ Exit the sketch.

Extruding to a Face

- ➢ Setup the **Extrusion** as shown in Figure 4.18 and accept these settings. Making sure that you choose the **Up To Surface** option as the end condition, select the face as shown as the surface and turn on the option to **Draft outward**.

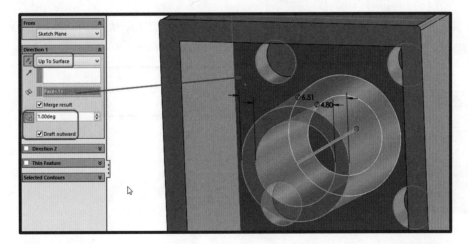

Figure 4.18

To see this being done, watch the video "Extrude to Face."

Adding the Material Type

➢ In the feature tree, right click on **Material** and select **Edit Material:**

➢ From the **Material** window, expand the **Plastics** and select **ABS**
➢ **Apply** this material to your part.
➢ **Close** the **Material** window

The feature tree should now show **ABS** as the material, and the part color should have changed to white.

Offset Plane

➢ Select the **Front Plane** from the feature tree

➢ From the **Features** tab, select **Reference Geometry, Plane**
➢ Set the **Distance** option at **3.9 mm** as shown in Figure 4.19.

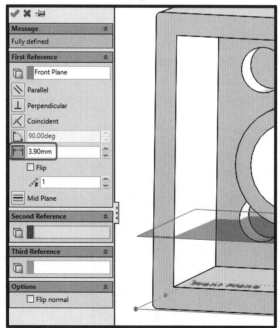

In SolidWorks 2014: If your plane is offset in the wrong direction (so that it will be created below the part), then check the box beside the **Flip** option. Your preview should look the same as shown in Figure 4.19.

In SolidWorks 2015: the appearance is slightly different as shown here:

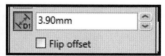

Figure 4.19

➢ Accept these selections

Chapter 4: LEGO Minions, Part 1—Designing 53

Working with a Cross Section

➢ If the plane you just created is not selected, then select it in the graphics area or in the feature tree (remember it will be highlighted in blue in the feature tree when selected).

➢ From the **View (Heads-Up)** toolbar, select **Section View**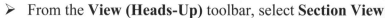

Your screen should look similar to Figure 4.20. If the larger portion of the part is not removed, then use the **Reverse Section Direction** option, shown outlined by the red rectangle, just below the #1 in Figure 4.20. You may want to use the **display with edges** display style.

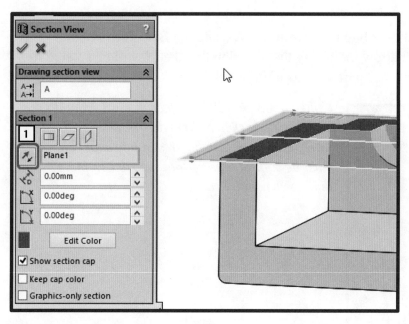

Figure 4.20

➢ Accept this selection ✓

Ribs

➢ Select **Plane1**.

➢ From the pop up window, select **Normal To**
➢ Select **Plane1** again.

➢ From the **Features** tab, select **Rib**

➤ Sketch two lines, one vertical (**0.3 mm**) and one horizontal (which ends coincident with the inside face), as shown in Figure 4.21. Be sure to fully define the sketch with relationships and dimensions.

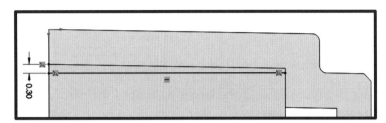

This image has been rotated 180 degrees.

Figure 4.21

➤ Exit the sketch.

➤ Following Figure 4.22, set the rib to be **0.6mm thick,** using the **Both Sides** option. There should be no draft, and the arrow indicating the direction that the rib will be created in should point towards the nearest part wall, as in Figure 4.22.

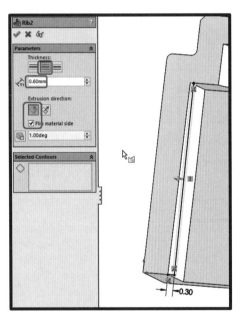

Unlike many SolidWorks features, ribs should not have a closed sketch. In order for a rib feature to work, the sketch can be almost any shape, but there must be a solid feature that the rib can merge into in the extrusion direction.

Figure 4.22

➤ Accept these selections.

➤ Turn the cross section off

➤ **Save your work.**

To see this being done, watch the video "**Rib Creation**."

Circular Pattern

- From the **View** menu
- Select **Temporary Axes** . This will show all cylindrical features axes.
- From the feature tree, select the **Rib** feature.
- From the **Features** tab, select the drop down arrow below **Linear Pattern** and then select

Circular Pattern

- As shown in Figure 4.23, select the **Temporary Axes** on the center of the part as the axes of rotation, set the rotation **angle** to **360 degrees**, set the **number of instances** to **4**, and turn on the check box for **Equal Spacing**. Accept these selections.

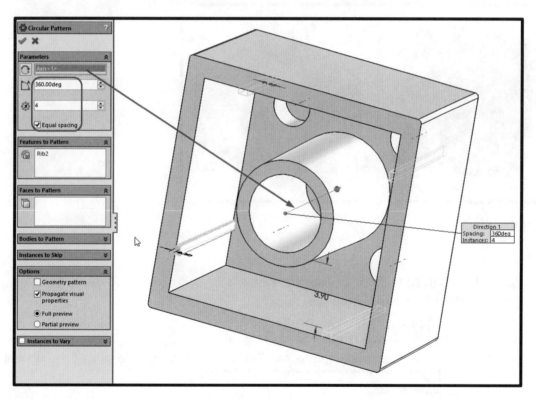

Figure 4.23

Parallel Plane Creation

- Select the front plane.
- Select **Reference Geometry, Plane.**
- Select the **Parallel** plane creation option

➤ Select the **Temporary Axes** located in the middle of the part as the **Second Reference**:

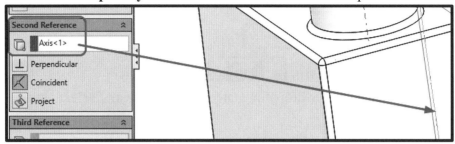

➤ Accept these selections.
➤ Create another **Parallel** plane by selecting the **Right Plane** first and using the same axes as the **Second Reference**.

Your screen should now look similar to Figure 4.24.

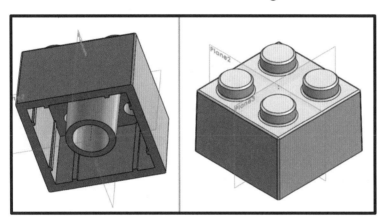

Figure 4.24

Mirror

➤ Rotate your part so that you can see the ribs.
➤ Select the **Rib** from the feature tree.
➤ Select **Mirror** from the **Features** tab.
➤ Select the newly created plane that is parallel to the **0.6 mm** wide side of the rib as the mirror plane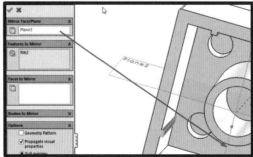
➤ Accept these selections.

- Move the **Circular Pattern** to the bottom of the feature tree

- Edit the **Circular Pattern** to include the **Mirror** feature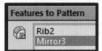

- **Save your work.**

*Remember, to edit a feature, select it in the feature tree and then select edit feature .
To add the mirror feature, click in the features to pattern a section and then select the feature
from the feature tree or from the part in the graphics area.*

Adding Color

- In the feature tree, right click on the part name and select the drop arrow beside the **Appearances** option (the beach ball) . Then select the part name.

*Appearances can be added to an entire part, a feature, a surface, or multiple surfaces.
Always set the parts material before adding any appearance, because the appearance will be
overridden by the default material appearance if it is added before the material.*

- From the color menu  on the left-hand side of the screen, select the **shiny** option and select the third red from the top.
- Accept these selections.

Your part should look similar to Figure 4.25.

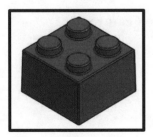

Figure 4.25

58   Chapter 4: LEGO Minions, Part 1—Designing

Practice

1) Create another block just like the previous one, but this time, change the **9.6 mm** height to **3.2 mm** . Save this as **ShortBlock_FirstName_LastName**. This block will have no internal features.

2) Create another block 7.9 x 7.9 x 9.6 with one round extrusion on top and no internal features other than the shell 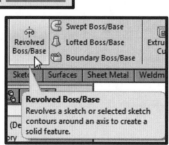. Save this as **OneBy_FirstName_LastName**.

Wheel Hub

Revolved Boss/Base

➢ Select the **Front Plane**

➢ Select **Revolved Boss/Base** from the **Features** tab.

➢ Using the **Line** sketching tool create the sketch as shown in Figure 4.26.
➢ Add the millimeter linear dimensions as shown in Figure 4.26.

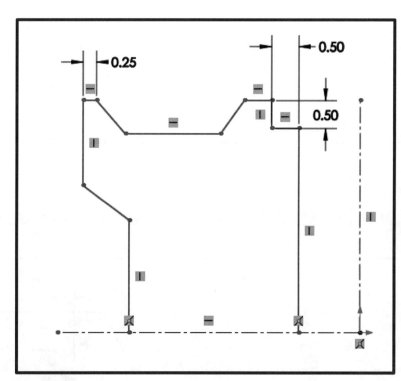

The vertical and the horizontal relationships should be automatically added as you create your sketch. Be certain to have the part origin as shown.

Notice how the center lines are black, except for the endpoints, indicating that they are fully defined. Center lines do not require their endpoints to be black in order for them to be fully defined, because their length is irrelevant.

Do not add any other relationships at this point.

Figure 4.26

Relationships

➢ While holding the **CTRL** key, select the two lines as shown in Figure 4.27.

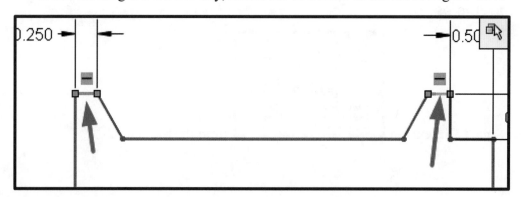

Figure 4.27

60 Chapter 4: LEGO Minions, Part 1—Designing

➢ Add the **Collinear** and the **Equal** relationships as shown in Figure 4.28.

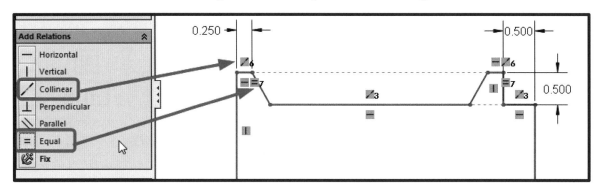

Figure 4.28

➢ Select the other two horizontal lines as shown in Figure 4.29 and again add the **Collinear** relationship.

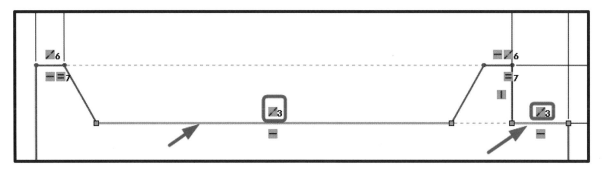

Figure 4.29

Dimensioning Angles

➢ Select **Smart Dimension** 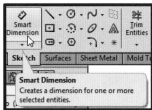 from the **Sketch** tab.

➢ Select the two lines as shown in Figure 4.30. *(The 0.25 dimension is shown as a reminder only and is not part of your sketch because you have the equal relationship.)*

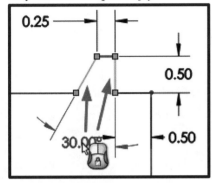

Figure 4.30

➢ Move your mouse down and click to place the dimension.

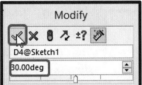

➢ Enter **30** as the angle and accept this selection.
➢ Select the two angled lines as shown in Figure 4.31 and make this dimension **60 degrees**.

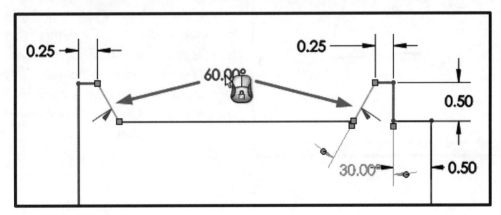

Figure 4.31

To see this being done, watch the video "**Dimensioning Angles**."

Diameter Dimensioning

➢ If **Smart Dimension** is not active, then turn it on.

Recall: if a command such as smart dimension is on, then it will have the appearance of a button that is pressed in and the color will be a darker gray than the rest of the commands. If it is not active, it will not appear to be depressed and will have the solid light gray appearance as the rest of the inactive commands.

➢ Select the object line and the center line as shown in Figure 4.32. Move your mouse to the right of the center line and click to place the dimension. Enter **2.5** as the hole size and accept this value.

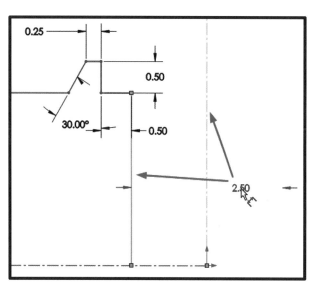

Figure 4.32

After selecting the object line and the center line, move your mouse to the right of the center line and then back to the left of the center line. You should notice that the dimension will double (indicating the features diameter) when the mouse is to the right of the center line and will display as the radius value when the mouse is on the left side of the line. Therefore, if a radius value is desired, click in between the two lines. If a diameter value is required, move your mouse past the center line and click to place the dimension.

After creating one diameter dimension, you should notice that the cursor has changed to show the mouse pointer, the dimension symbol, a center line, and a capital letter D, indicating that diameter dimensioning is now active. This feature will stay active until you exit the dimensioning command or until you press the **ESC** key once to set it back to the standard dimensioning mode. As long as the diameter dimensioning is active, each object that you select to dimension will be given a diameter dimension.

This technique can also be used for dimensioning non-diameter features when only half of the part has been drawn; the rest will be mirrored through a center line after dimensioning.

Continue using the diameter dimensions until your sketch is fully defined and you have dimensioned all of the features as shown in Figure 4.33. Do not add any other dimensions, if your sketch is not **Fully Defined** Fully Defined, as this means that you are missing sketch relationships. First, find the missing relationships and add them. (*You will need to press the ESC key on the keyboard to end diameter dimensioning to the vertical center line and to start it from the horizontal center line.*)

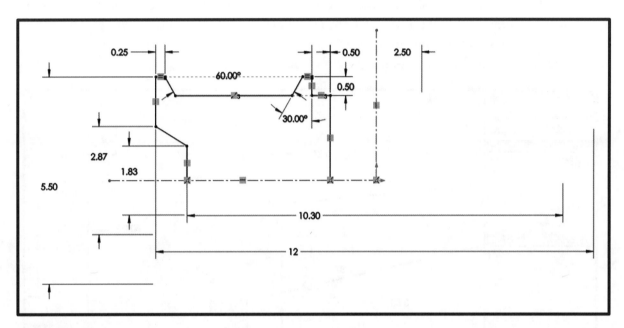

Figure 4.33

Remember, when a sketch is fully defined, all of the object lines will be black and SolidWorks will display "Fully Defined" in the status bar at the bottom of the screen. The end points of the center lines can be blue, and the sketch still be fully defined, because the length of a center line is irrelevant.

An under defined sketch will also show a minus sign (-) Sketch1 beside its name in the feature tree.

If you think your sketch is fully defined but SolidWorks is displaying this indication that it is **Under Defined** Under Defined, then you most likely have accidentally added a very small line. Look for the blue end point, click and hold on it, then drag it out so that it is long enough to see and delete it.

To see this being done, watch the video "**Diameter Dimensioning**."

When dimensioning, try to place the dimensions where you would like them to be in the part drawing, because when you place them in a sketch, this is where they will initially appear in the drawing.

Sketch Mirror

> Select **Mirror Entities** from the **Sketch** tab. SolidWorks will open the mirror command with the **Entities to mirror** field highlighted, thus allowing you to immediately start selecting entities
> Select all of your solid object lines as shown in Figure 4.34.

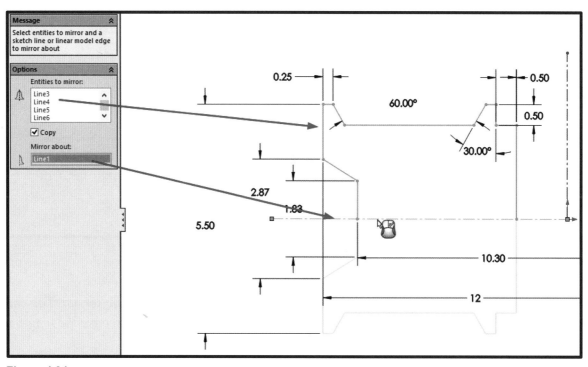

Figure 4.34

> Click in the Mirror about box then select the horizontal center line as the line to mirror the entities. Do not uncheck the **Copy** option.
> Accept these selections.

Again if you do not move your mouse after selecting the mirror line, you can right click to accept the selections and exit the command.

Note that you can window select the entities to mirror or select each line individually. After selecting these entities, if you do not move your mouse, a picture of a mouse with a blue return arrow on the right button will appear. This is SolidWorks telling you that you can right click to end your selection of entities to mirror and then immediately select the line to mirror.

Chapter 4: LEGO Minions, Part 1—Designing   65

To see this being done, watch the video "Sketch Mirror."

➢ Exit the sketch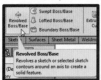

If you have not exited the sketch and re-entered it, then you will be taken to the revolve boss/base feature window, however; if you have exited the sketch and re-entered it, then when you exit the sketch, the revolve feature window will not be active.

If this is the case, then select the sketch from the feature tree and then select **Revolved Boss/Base** 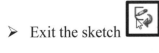 from the **Features** tab.

➢ Select the vertical center line as the **Axis of Revolution** as shown in Figure 4.35.

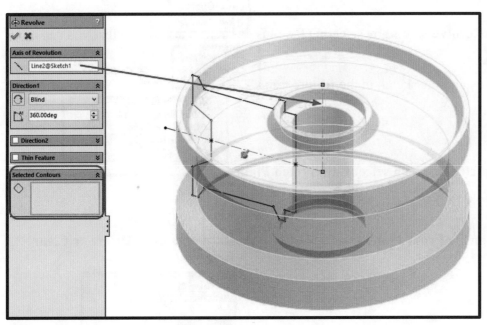

*If you do not see a preview of your part in yellow, then you will need to click inside the **Selected Contours** box and then click inside your sketch above and below the center line.*

Figure 4.35

➢ Accept these selections

66 Chapter 4: LEGO Minions, Part 1—Designing

> - Set your material to ABS. Make the part color a light gray.
> - **Save your work** as **Hub_FirstName_LastName**.

Your part should look similar to Figure 4.36.

Figure 4.36

Switching the Sketch Plane

After you create a feature you may realize that it is in the wrong sketch plane. Don't panic, as this is an easy fix.

> - Expand the **Revolve** feature by selecting the plus  sign so that you can now see the feature and the sketch in the feature tree

> - Select the **Sketch** and then select **Edit Sketch Plane**

> - Right click on the **Front Plane** 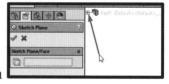 and select **Delete or Clear Selection**.

> - Expand the feature tree by clicking on the plus sign

> - Select the **Top Plane** 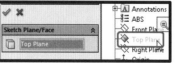 and accept these selections.

To see this being done, watch the video "<u>Switching Sketch Planes</u>."

Chapter 4: LEGO Minions, Part 1—Designing **67**

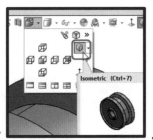

> Select the **Isometric View** .

Your part should now look similar to Figure 4.37.

Figure 4.37

Extruded Cut

> Select the front face of the part as shown in Figure 4.38 and apply **Normal To** it.

Figure 4.38

68 Chapter 4: LEGO Minions, Part 1—Designing

➢ With the face still selected *(it will still be blue in color)* select **Extruded Cut**: from the **Feature** tab. *(Click in the graphics area but not on the part to remove the blue color from the part face.)*

➢ Sketch the entities as shown in Figure 4.39. Add all of the relationships and dimensions as shown.

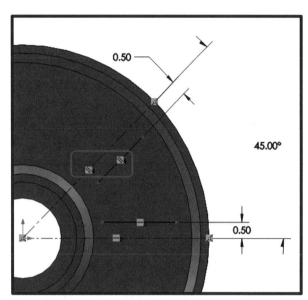

The relationship shown outlined in red is Parallel

Figure 4.39

Offset Entities

➢ Select **Offset Entities**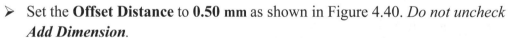
➢ Set the **Offset Distance** to **0.50 mm** as shown in Figure 4.40. *Do not uncheck Add Dimension*.
➢ Select the edge of the part as shown in Figure 4.40 as the entity to offset.

If your circle is not larger than the selected circle then check the box beside Reverse.

Figure 4.40

➢ Accept these selections.
➢ Offset the larger diameter edge that is on the same face, towards the center by 0.5.

Your screen should now look similar to Figure 4.41.

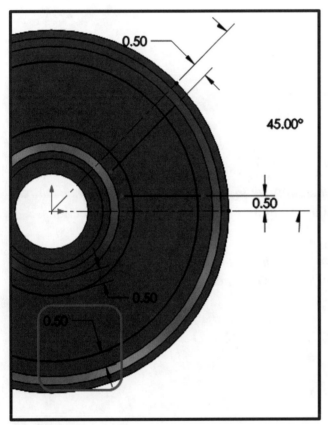

If your lines do not cross the two offset circles, then click on the blue end points and drag them until they look similar to the Figure 4.41 before proceeding to the next step.

The coincident relationship of the center lines to the origin may be deleted when you offset the part edge. If these lines turn blue, you will need to move the end point of the center line off of the origin and add the coincident relationship again.

Figure 4.41

To see this being done, watch the video "**Offsetting Sketch Entities**."

Trimming Entities

➢ Select **trim Entities** . *Power Trim will be active by default. Do not change this.*

➢ Position your mouse on your sketch as shown in Figure 4.42.

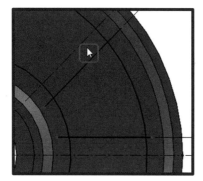

Figure 4.42

➢ Click and drag your mouse through the sections of the lines and arcs that you want removed. **Power trim** *will remove the entity up to the point it crosses another entity.*
➢ Trim your sketch until it looks similar to Figure 4.43 and accept these selections.

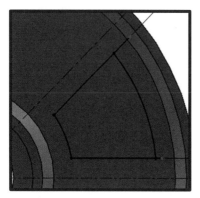

Your dimensions may become deleted after you trim, therefore you may need to add them again.

Figure 4.43

To see this being done, watch the video "**Trimming Entities**."

Merge Relationship

➢ Select one of the lines blue colored end points.
➢ Hold the **CTRL** key and select the nearest black arc end point.
➢ Add the **Merge** relation
➢ Repeat this for the other endpoint.

Sketch Fillet

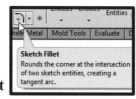

➢ Select **Sketch Fillet**

Chapter 4: LEGO Minions, Part 1—Designing 71

➢ Enter **0.5**  as the size.
➢ Select the four corners of the sketch as shown in Figure 4.44 or select the line and then the arc near each corner and accept these selections.
➢ Exit the command. Your sketch should look similar to Figure 4.45.

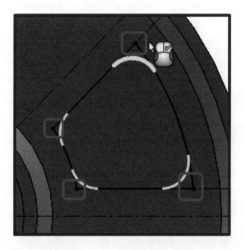

Figure 4.44

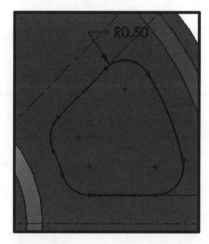

Figure 4.45

➢ Exit the sketch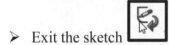

➢ Set the end condition for the cut to **Through All** as shown in Figure 4.46 and accept these selections.

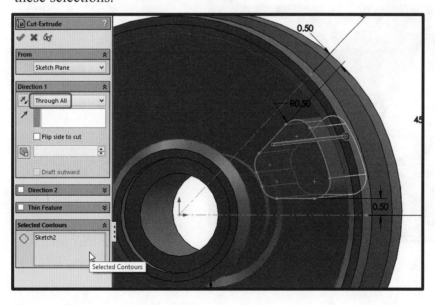

Figure 4.46

➢ **Save your work.**

➢ Add a **0.1 mm** feature fillet to each end of the cut extrude. Your part should look similar to Figure 4.47.

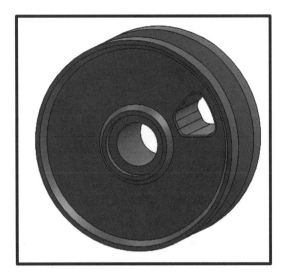

Figure 4.47

When you fillet your part, try selecting just one edge and not moving your mouse. A window should appear as shown in Figure 4.48. Move your mouse over each option presented and pay attention to how the fillet feature changes.

Figure 4.48

Circular Sketch Pattern / Edit a Sketch

- Click on your **Cut Extrude** feature and select **Edit Sketch**
- Look **Normal To** your sketch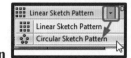
- Select the drop down arrow beside **Linear Sketch Pattern** and then select **Circular**

Sketch Pattern
- Select the origin as the **Rotation Point**, set the number of instances to **8**, set the **Rotation Angle** to **360**, and turn on the check box for **Equal Spacing**.
- Select the object lines as the entities to rotate as shown in Figure 4.49 and accept these selections.

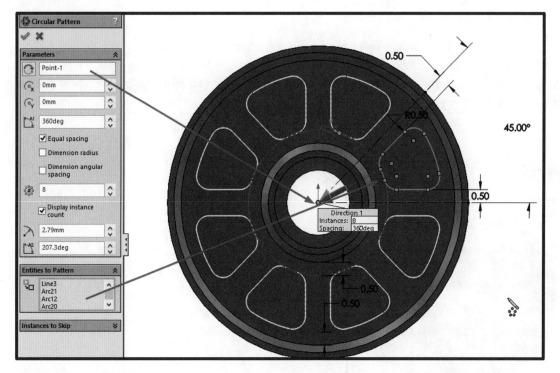

Figure 4.49

- Exit the sketch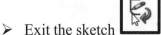

To see this being done, watch the video "**Circular Sketch Pattern**."

74 Chapter 4: LEGO Minions, Part 1—Designing

- Edit the fillet to now include all edges of the part.
- Window select the entire part. Your part should now look similar to Figure 4.50.

Figure 4.50

- **Save your work.**

So how do you know when you should use a feature or a sketch pattern? As you become a more experienced user of SolidWorks this will become self-evident. For now try to always use a feature pattern. As soon as we cover design configurations in more detail, you will understand why feature patterns are preferred.

Tire

Figure 4.51

Create the tire, minus the tread as shown in Figure 4.51 above, using the following information:

- For material, specify BUTYL rubber.
- Using only the dimensions given and relationships as required to fully define all of your sketches, start in the front plane and create a revolved boss as your base feature, as shown in figure 4.52.

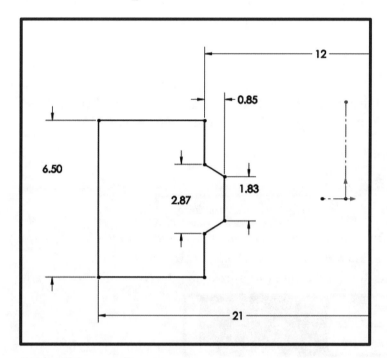

Figure 4.52

Symmetry Relationship

- Create the sketch as shown in Figure 4.52 above, including the dimensions shown.
- Apply the obvious horizontal, equal, and vertical relationships.
- Select the lower horizontal line, hold the **CTRL** key, select the top horizontal line, and while still holding the **CTRL** key, select the horizontal center line.
- Add the **Symmetric** [Symmetric] relationship.
- Add the **Symmetric** relationship to the two angled lines as well to fully define your sketch.

To see this being done, watch the video "Symmetrical Relationship."

Chapter 4: Lego Minions, Part 1—Designing

Customizing Commands

All of the commands are available from the drop down menu across the top. However, most people tend to use the ribbon bar, and this can easily be customized to any user's preferences.

*Earlier, you added the **Dynamic Mirror** command to the **Sketch** ribbon. Now, you will use it. If this command is not in your sketch ribbon bar, then add it again now.*

Tread Extrusion

- Select the top plane and then select **Extruded Cut** from the **Feature** tab.
- Create a vertical center line from the origin to the outside diameter edge.

Dynamic Mirror

- Select **Dynamic Mirror Entities** *(This will mirror all entities as you create them, the command can also be found from the top drop down menu by selecting **Tools, Sketch Tools,** and **Dynamic Mirror**.)*
- Select the vertical center line as the entity to mirror about. (Notice the hash marks on the center line . This image has been rotated 180 degrees to save space here.)
- In the top plane, create the sketch as shown in Figure 4.53. Add all of the relationships as shown.

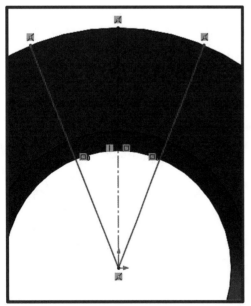

Figure 4.53

The two object lines are symmetrical about the center line and coincident with the origin and the outside diameter edge.

The center line is vertical. You should notice that as soon as you click to place the second end point of the line, it is mirrored through the center line.

Chapter 4: LEGO Minions, Part 1—Designing **77**

Center Point Arc

- Select **Center Point Arc**
- Click on the part origin to lock the arc center.
- Move your mouse up and click on the center line.
- Move your mouse to the right and click on the object line near the outer diameter.
- Exit the command.
- Dimension the arc **10 mm**. Your screen should look similar to Figure 4.54.

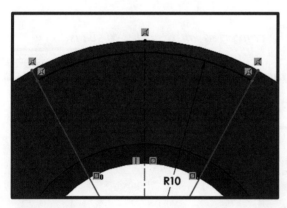

Figure 4.54

- Turn off the **Dynamic Mirror Entities** by selecting it again.

To see how to add the dynamic mirror command to the ribbon bar and how to use it watch the video "**Counter Block Cut Top Slot**."

For an explanation of how to create arcs, watch the video "**SolidWorks Sketch Arc**."

Convert Entities

- Select **Convert Entities**
- Select the outside edge of your part.
- Accept these selections

➤ Use **Power Trim** to make your sketch look similar to Figure 4.55 (trim top/bottom).

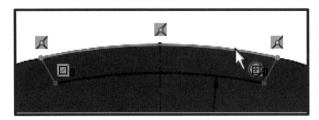

Figure 4.55

When converting entities on a part such as this, it is often difficult to see the converted edge. However, after you trim the geometry, if you position your mouse near the edge, it will change color to orange. Notice how the converted circle has been trimmed.

Another option which can be used to make the converted entity easier to see is to right click on the feature in the feature tree and select **Change Transparency**  *as shown in Figure 4.56.*

Figure 4.56

You also can select the Hide *option as shown in Figure 4.57.*

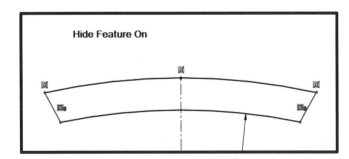

Figure 4.57

Chapter 4: LEGO Minions, Part 1—Designing

Angular Dimensioning using Points

- Add a **Coincident** [Coincident] relationship between one of the angled object lines and the origin. *(Once these two lines were trimmed, the Coincident relationship to the origin that was initally present was deleted.)*
- Select **Smart Dimension**.
- Select the converted arcs right endpoint, then the origin, and then the converted arcs left endpoint, as shown in Figure 4.58 Then place the dimension above the arc. Make the angle **13 degrees**.
- Add a horizontal center line as shown in Figure 4.59.
- Mirror the sketch through this center line, so that your sketch now looks like Figure 4.59.

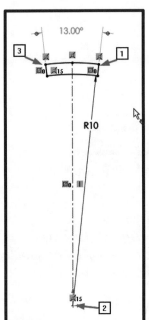

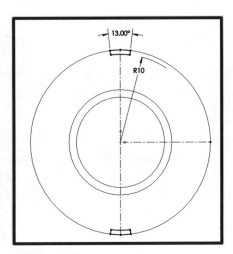

The actual order that the three points are selected in does not matter. You can always select the two lines to create the angular dimension.

Figure 4.58   **Figure 4.59**

- Exit the sketch

Through All Extruded Cut

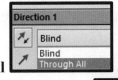

- Select **Through All** as the **End Condition** for the extrusion.

- In the **Selected Contours** section, select only the top contour.

80 Chapter 4: LEGO Minions, Part 1—Designing

➤ Accept this to create the cut. Your part should now look similar to Figure 4.60.

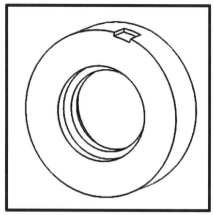
Figure 4.60

Note that the part displayed has had **Hidden Lines Removed** *for visual clarity*

➤ **Save your work** as **YourName_Wheel**.

Sharing a Sketch

➤ Expand the **Cut-Extrude** feature so you can see the sketch
➤ Expand the **Revolve** feature as well. Your feature tree should look similar to Figure 4.61.

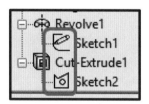
Figure 4.61

The image beside Sketch1 is displayed when your sketch contains one closed object. The image beside Sketch2 is displayed when there is more than one closed object. As soon as you share your second sketch, you will notice it now has a picture of a hand holding the sketch.

➤ Select the sketch for the **Cut-Extrude**
➤ From the **Features** tab, select **Extruded Cut**.
➤ Set the end condition to **Through All**.
➤ Reverse the direction of the cut 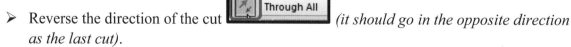 *(it should go in the opposite direction as the last cut)*.
➤ Right click in the **Selected Contours** section and select **Clear Selections**.
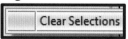

➢ Select the lower contour and accept these selections. Your part should now look similar to Figure 4.62.

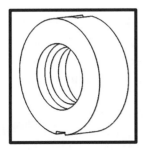

Figure 4.62

To see this being done, watch the video "Share Sketch."

Re-Naming Features

➢ In the feature tree, right click on your first cut extrude feature and select **Feature Properties**
➢ In the **Feature Properties** window, click in the **Name** cell and enter: **Top Tread Cut**.
➢ Select **OK**
➢ Hold the **Shift** key and select both **Cut-Extrude** Features in the feature tree.

➢ Select the **Circular Pattern** from the feature tab.
➢ Select the outside diameter face as the pattern axis. *(You can also select a temporary axis or a circular edge or any diameter face as the pattern axis.)*
➢ The pattern has thirteen equally spaced instances.
➢ Accept these selections. Your part should look similar to Figure 4.63.

Figure 4.63

➢ **Save your work.**

Design Configurations—Parametric Modeling

Design configurations allow you to create one part that can have multiple variations in its sizes and feature. This is known as parametric modeling, since the design parameters can be changed to create multiple variations of a part. When creating design configurations, you are not required to rename features and dimensions, but it will help to keep things organized and easier to work with. The configurations can be created solely within SolidWorks or in a spreadsheet, which can be linked to your part or assembly file.

- Open your **Tall Block.**
- Select the **Configuration Manager**  tab found at the top of the design tree.

- You will now see your part name and the one configuration listed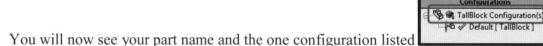
- Right click on the part name**, not on the configuration**, and select **Add Configuration:**

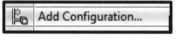

- Name this configuration **2x3** 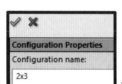 and accept this name.
- You **may** receive the warning about the part color as shown in Figure 4.64. If you do receive this warning, select **Yes**.

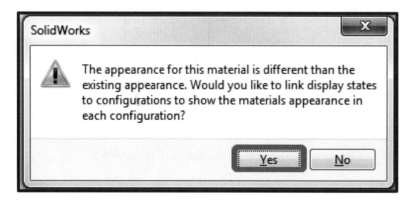

Figure 4.64

➢ Your configuration tree should look like Figure 4.65, with the new configuration highlighted. *It will be an exact copy of the initial part at this point in time.*

If your new configuration is not as shown here but instead is a new branch off of and below the original configuration, then delete it and start again by selecting the part name at the very top of the list as shown in Figure 4.65.

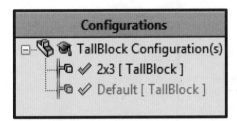

Figure 4.65

➢ Add the following configurations:
  o 2x4
  o 2x5
  o 2x6
  o 2x7
  o 2x8
  o 2x9
  o 2x10

until your configuration tree looks like Figure 4.66.

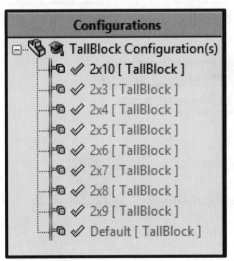

Figure 4.66

Note: Each time you add a configuration, you must start by right clicking on the part name at the top.

To see this being done, watch the video "**Creating Design Configurations**."

Naming Dimensions

➢ Double click on your part in the graphics area. The dimensions will display around your part as shown in Figure 4.67.

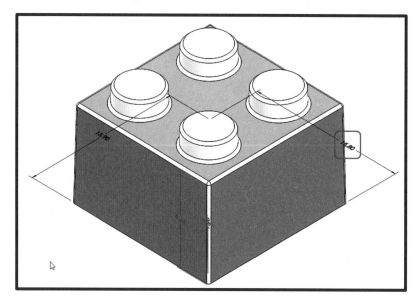

Figure 4.67

➢ Select the **15.8** length dimensions.

➢ On the left side of your screen, you will see the **Primary Value** for this dimension. It is currently called **D1@Sketch1.** (*Your primary value may be a different number.*)

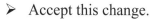

➢ Rename this value to **Length@Sketch1**
➢ Accept this change.

To see this being done, watch the video "<u>Naming Dimensions</u>."

Chapter 4: LEGO Minions, Part 1—Designing

Configuring External Dimensions

> Double click on your part again to show the dimensions.
> Right click on 15.8 dimension and select **Configure Dimension**
> Enter the values in the table as shown in Figure 4.68. Accept these selections.

Configuration Name	Sketch1 Length
2x10	79.80mm
2x3	23.90mm
2x4	31.80mm
2x5	39.80mm
2x6	47.80mm
2x7	55.80mm
2x8	63.80mm
2x9	71.80mm
Default	15.80mm
< Creates a new configuration. >	

Figure 4.68

*You should notice that as soon as you accept these values and leave the **Modify Configurations** window that the part updates to the length of the currently selected configuration:*

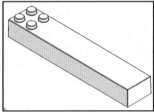

To see this being done, watch the video "**Configure Dimension**."

If you receive an error message, just close it for now. Shortly, you will be correcting this error. Often when creating design configurations, unanticipated problems present themselves.

Selecting a Configuration

➤ Select the **Configuration Manager** tab
➤ Double click on the name of a configuration to select it.

At this point in time, when you select a configuration you should receive two error messages similar to that shown in Figure 4.69. **Rebuild** your part and **Close** the **What's Wrong** message.

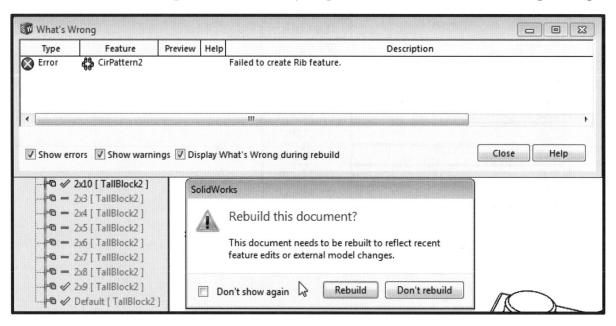

Figure 4.69

➤ Select the Feature Manager Design Tree tab.

➤ You should notice that your circular pattern has a red "**X**" on the feature in the design tree

This is telling you that the pattern cannot work for this configuration, so you will need to fix it to a style that works for all configurations. If you look at the bottom of your part, you should also notice that the third boss extrude feature is also out of position as shown in Figure 4.70.

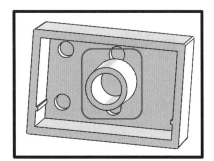

Figure 4.70

Chapter 4: LEGO Minions, Part 1—Designing

Repairing Issues

➢ From the **Configuration Manager** tab, select (double click on) your **Default** configuration

➢ Go back to the **Feature Manager Design Tree**.

The circular pattern will no longer work.

➢ Right click on the **CirPattern** feature and select **Delete**

➢ Confirm the deletion of this feature

➢ Rebuild and save your part.

➢ Right click on your third **Boss-Extrude** feature and select **Edit Sketch**

➢ Select and then delete the two **Midpoint** relationships that are being used to lock this feature to the center of the part.

To see this being done, watch the video "**Deleting Constraints**."

➢ Re-dimension this feature from the part origin as shown in Figure 4.71, exit the sketch, and save your changes.

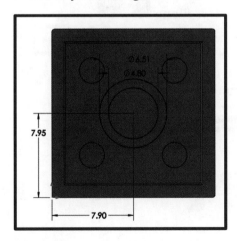

Figure 4.71

Edit Plane 2

➢ Delete the **Second Reference** and replace it with the part edge as shown in Figure 4.72.

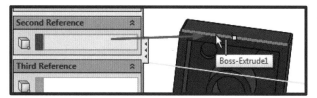

This edge is opposite the front plane.

Figure 4.72

Edit Plane 3

➢ Delete the **Second Reference** and replace it with the part edge as shown in Figure 4.73.

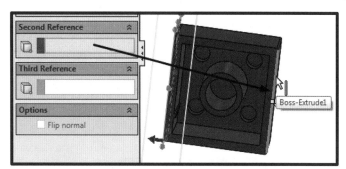

This edge is opposite the right plane.

Figure 4.73

➢ Move the **Rollback Bar** to above your **Mirror** feature

Mid-Plane Plane

➢ Select the **Front** plane.

➢ From the **Features** tab, select **Reference Geometry, Plane**

➢ Near the bottom of the **First Reference** section, select **Mid Plane**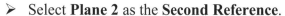
➢ Select **Plane 2** as the **Second Reference**.
➢ Accept these selections to create a new plane on the center of the part.
➢ Rename this plane **Front Midplane**

To see this being done, watch the video "<u>Mid Plane Creation</u>."

- Create an offset plane **7.9 mm** (this should be in the center of the part) from the **Right Plane** and name it **Right Midplane**.
- Rename your third **Boss-Extrude** to **Internal Extrude**.

- Create an offset plane 3.9mm from the **Right Plane** (*this plane should go through the part not through the air beside the part*). Name this plane **Right Rib**.

Creating a New View Orientation

- Rotate your part by pressing in the middle mouse wheel and holding it while moving the mouse around. Stop rotating your part when it is in an orientation similar to that shown in Figure 4.74.

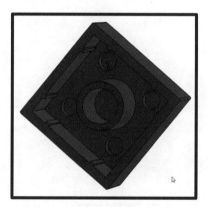

Figure 4.74

- Press the **Space Bar** on your keyboard.

- Select **New View**  from the **Orientation** window.

- Name this view **Bottom1** and select **OK**
- Close the **Orientation** window.

*This view will now be a selectable view every time you select the view **Orientation**.*

To see this being done, watch the video "New View Orientation."

Copying a Sketch

> Expand your rib feature. Right click on the sketch consumed by your rib and select 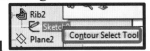 **Contour Select Tool**
> Press **CTRL + C** on your keyboard.
> Right click on the sketch again and select **End Select Contours**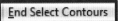
> Select the **Right Rib** plane.
> Select **Rib** from the **Features** tab.
> Look **Normal To** ⊥ your sketch plane.
> Press **CTRL** and **V** together on the keyboard to paste the sketch.

To see this being done, watch the video "Copy Sketch."

> To the horizontal line, add the **Horizontal Relationship** ▭ ; to the vertical line, add the **Vertical Relationship**. (*In SolidWorks 2015:* The dimension and the relationships will also be copied. Delete the **0.3 mm** dimension.)

> Add the **Coincident** ⟨Coincident⟩ relationship to the open end point of the horizontal line and the origin.

> From the **Standard Views** toolbar, select the **Hidden Lines Visible** from the **Display**  **Style** option

➢ Add the **Coincident** relationship to the open end of the vertical line and the lower inside edge of the part as shown in Figure 4.75.

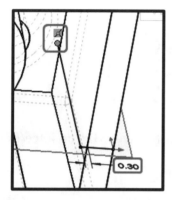

Figure 4.75

Depending on how you select your entities for this relationship, you may not see the **Coincident** *option, rather you may see the* **Pierce** *relationship. If this is the case, it is fine, select the* **Pierce** *relationship*.

➢ Add the **0.3 mm** dimension from the inside edge of the part to the end of the horizontal line.
➢ Make the rib the same width (**0.6**) and type as the first rib (midplane).

➢ Set your display back to **Shaded With Edges**
➢ Add two **Linear Pattern Features**, one for each rib. The pattern should produce two ribs **8.1 mm** apart. Your part should now look similar to Figure 4.76.

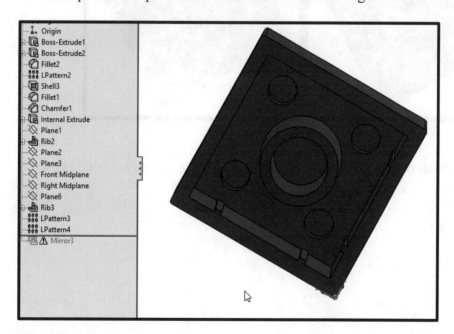

Figure 4.76

- ➢ Move the **Rollback Bar** to the bottom of the feature tree.
- ➢ Edit the **Mirror** feature

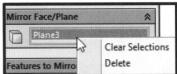

- ➢ Right click in the **Mirror Face/Plane** and select **Clear Selections** or **Delete**.
- ➢ Expand the feature tree and select the **Right Midplane** as the new **Mirror Face/Plane**

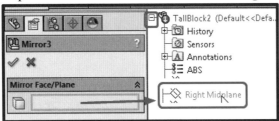

- ➢ Add the linear pattern of this rib to the **Features to Mirror**
- ➢ Accept these selections. Your part should look similar to the left side of Figure 4.77.

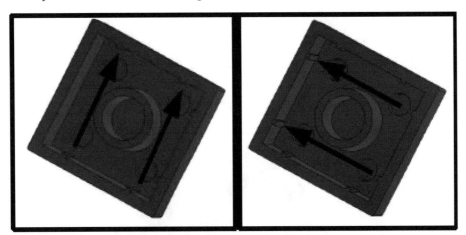

Figure 4.77

- ➢ Add a similar **Mirror** feature for the other rib and linear pattern so that your work now looks similar to the right side of Figure 4.77.

- ➢ **Save your work.**

Configuring Internal Features

➢ Select your **2x3** configuration by double clicking on it. Your screen should look similar to Figure 4.78.

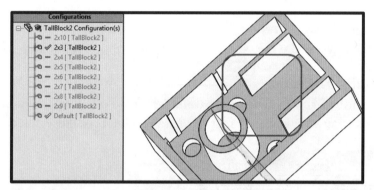

The one set of mirrored ribs are too long because you have not yet moved the mirror plane.

Figure 4.78

➢ In the **Feature Manager Design Tree,** double click on your **Right Midplane**.
➢ Right click on the **7.9** dimension and select **Configure Dimension**

Configure all of the dimensions as shown in Figure 4.79. Accept these selections.

Configuration Name	Right Midplane D1
2x10	39.90mm
2x3	11.95mm
2x4	15.90mm
2x5	19.90mm
2x6	23.90mm
2x7	27.90mm
2x8	31.90mm
2x9	35.90mm
Default	7.90mm
< Creates a new configuration. >	

Check each configuration as you did before, all configurations should update without any errors. They will not be complete, but they should be error free.

Figure 4.79

If some of the features are suppressed *when the configuration is selected, then right click on feature and select Unsuppress*

➢ **Save your work.**

94 Chapter 4: LEGO Minions, Part 1—Designing

Suppressing/Unsuppressing Features

It is very likely as you verify your configurations are working that some of the features will be suppressed. Suppressed features appear grayed out 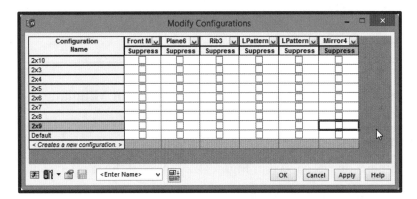. *You do not want this as of yet.*

> ➢ Select the **Feature Manager Design Tree**.
> ➢ Hold the **CTRL** key on the keyboard while selecting all of the suppressed features.
> ➢ Right click and select **Configure Feature**
> ➢ Uncheck all of the boxes in the **Modify Configurations** window so that your window looks like Figure 4.80 (*you may have more or less features*) and accept these selections.

Figure 4.80

Configuring Features

> ➢ Rename your first linear pattern feature **TopRounds**.
> ➢ Double click on this feature and all of the dimensions will show on the screen.
> ➢ Right click on the instance number and select **Configure Dimension**:

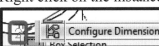

> ➢ If you get an error message about SolidWorks being unable to find the properties file, select **OK**.
> ➢ Set up the values as shown in Figure 4.81 and accept these selections.

Configuration Name	TopRounds D1
2x10	10
2x3	3
2x4	4
2x5	5
2x6	5
2x7	5
2x8	5
2x9	9
Default	2
< Creates a new configuration. >	

Every time you configure a new feature or dimension, you should check that each configuration is correct by double clicking on it.

*If there are any rebuild issues, correct them before continuing and **save your work**.*

Figure 4.81

By now, you should have noticed two issues with your configurations:
1) Only the original configuration has the assigned color.
2) The chamfer and the fillets are not part of the pattern on the new configurations.

Configuring the Color

> Select the default display state.

> Select the **Display Manager** tab

> Right click on the **Red ball** 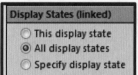 and select **Edit Appearance**.
> In the **Display States (linked)** section, select the **Specify display state** radio button and then select the **All display states** radio button and accept these selections (*check that all configurations now have the same appearance*).
> Configure the appropriate rib pattern so that your part has all of the ribs as required. Your 2x10 configuration should look similar to Figure 4.82.

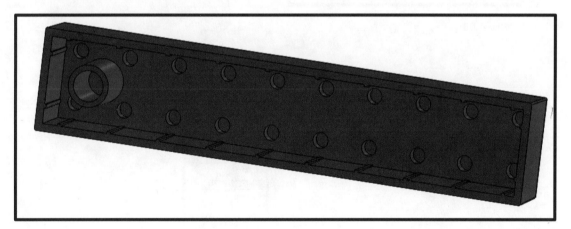

Figure 4.82

> Make your 2x10 configuration active.
> Add a linear pattern with a spacing of 8.0 for your **Internal Extrude** feature and configure this for all configurations. Name this pattern **InternalExtPattern**.

*All of your configurations should now work **except** for your default configuration.*

Chapter 4: LEGO Minions, Part 1—Designing

- Select your default configuration, rebuild your part, and close the error message.
- Select the **Feature** tab.
- Right click on the **InternalExtPattern** and select **Configure Feature**.
- Check the **Suppress** box beside the **Default** feature and accept these changes.
- Verify all configurations work. (*You may need to unsuppress other configurations.*)
- **Save your work.**

Revolve Cut

- Select your 2x3 configuration.

Axis Creation

- Select the face as shown in Figure 4.83.

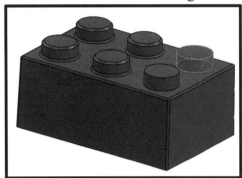

Figure 4.83

- Select **Reference Geometry** and then select **Axis**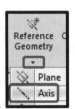

- Accept these selections to create the **Axis**

- Select **View, Temporary Axes**
- Select **Reference Geometry, Plane.**

Chapter 4: LEGO Minions, Part 1—Designing **97**

➢ Select the two axes as shown in Figure 4.84 and accept these selections.

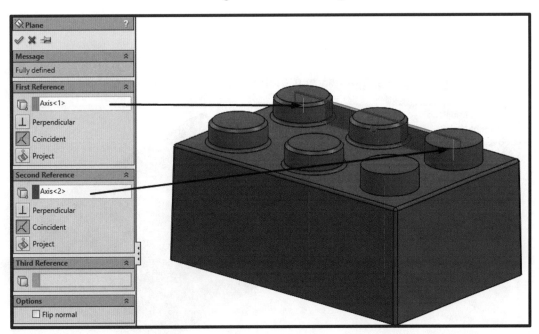

Figure 4.84

➢ Select your newly created plane.
➢ Select **Revolved Cut** from the **Feature** tab.
➢ Create the sketch as shown in Figure 4.85. *Note the line that appears vertical is not. It should be created using the convert entities command and converting the edge of the boss. You will either need to trim the converted entities or move their end points.*

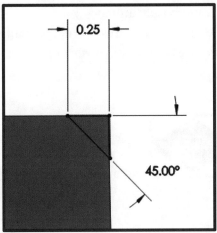

Figure 4.85

98 Chapter 4: LEGO Minions, Part 1—Designing

➢ Exit the sketch and select the axis you created earlier as the axis of revolution for the cut feature . *The direction of the revolution does not matter as it will be a full 360 degrees.*

➢ Accept these selections. Your screen should look similar to Figure 4.86.

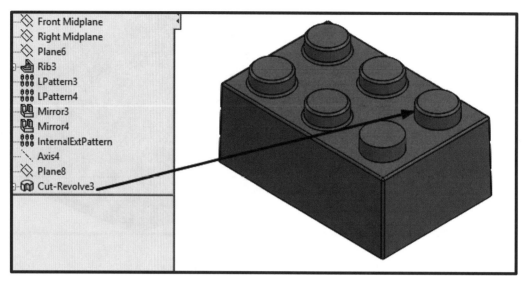

Figure 4.86

To see this being done, watch the video "**Revolve Cut Feature**."

➢ Using the same plane, create a fillet using the revolve boss feature.
Use the pattern and/or the mirror features to add these two features correctly to all part configurations. *You may need to suppress or unsuppress features as part of your configurations.*

➢ **Save your work.**

➢ For your **ShortBlock_FirstName_LastName** add a **4x6** (47.8 x 31.6) configuration.

➢ For your **OneBy_FirstName_LastName** create the following configurations:
  o **1x2** (15.9 x 7.9)
  o **1x3** (23.9 x 7.9)
  o **1x4** (31.9 x 7.9)

Derived Configurations

- Open your **OneBy_FirstName_LastName** part.
- Open the **1x2** configuration. *(Recall: to open a configuration you need to go to the configuration manager and double click on the desired configuration.)*
- Right click on the **1x2** configuration and select **Add Derived Configuration**  *(**Do not right click on the part name** as you did to make the earlier configurations. This configuration will be a variation of only this 1x2 part.)*
- Name this configuration **1x2 Angle**.
- Select the front plane of your part and select **Extruded Cut**
- Create the sketch as shown in Figure 4.87.

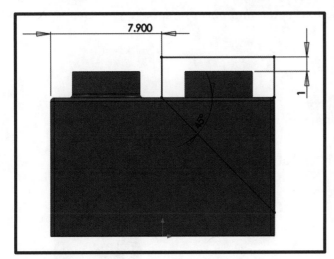

When you dimension features by default, these dimensions will appear in the part drawing. However, there are times when dimensions are required to fully define a sketch but are not wanted in the part drawing. By right clicking, we can set them to not display in the part drawing.

Figure 4.87

Mark for Drawing

- Right click on the "1" dimension and select Mark for Drawing
  (This will uncheck this option and stop the dimension from appearing in the part drawing.

  You should also notice the dimensions color changes to a shade of purple , a visual cue that it is not marked for the drawing).
- Exit the sketch.
- Set the end condition for the extrusion to **Through all** and accept these selections.

Your part should now look similar to Figure 4.88.

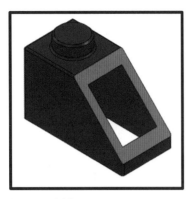

Figure 4.88

Do not create this cut above the shell feature or the fillet. Instead, leave it at the end of the feature tree. If you place this feature above the fillet, you will receive a warning that the fillet feature has problems. This would be due to the fact that part of the edge you used to create the fillet on no longer exists.

➤ Select the cut face and apply **Normal To** it.
➤ Select **Extruded Boss/Base**.
➤ Click beside the part but not on it so that the face is no longer blue.

➤ Select **Convert Entities**

➤ Right click on one of the faces edges and choose **Select Loop**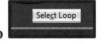

If the wrong face is selected as shown in Figure 4.89, then click on the yellow arrow to switch the loop to the outer edge of the originally selected face. This will create a fully defined sketch on the surface. If you alter the original part size, this sketch will also be altered, because it will always be on the part edge regardless of the feature size.

Figure 4.89

➢ Accept this selection
➢ Set the direction of the extrusion into the part.
➢ Make the distance **1.2 mm** and accept these selections as shown in Figure 4.90.

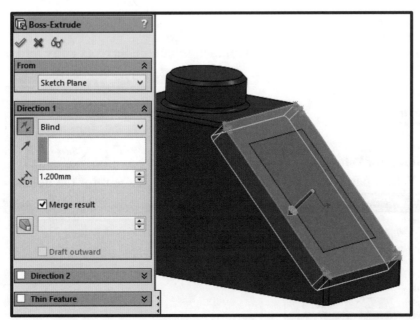

*Do not uncheck the **Merge result** box as this will cause the creation of two or more solids in one part.*

Most of the time, we are trying to create just one solid part.

Figure 4.90

Add a 0.25 fillet to this face.

➢ **Save your work.**

Your part should now look similar to Figure 4.91.

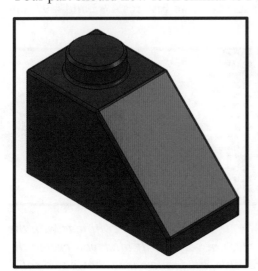

Figure 4.91

Save as Copy

- Open your **TallBlock_FirstName_LastName** part.
- Select the default configuration.
- Select **File, Save As**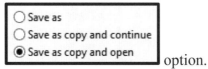
- In the lower left-hand corner of the **Save As** window, select the **Save as copy and open** option.
- Name this copy **BlockForWheel** and **Save** it.
- In your **BlockForWheel** part, select all of the configurations, except for the default and delete them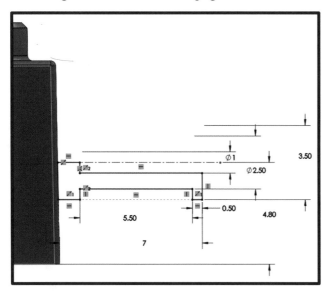
- In the **Confirm Delete** window, select **Yes to All**
- In the feature tree, delete all of the suppressed features.
- In the **Front Midplane** create a **Revolved** Boss as shown in Figures 4.92 and 4.93, and in Figure 4.94 on the next page.

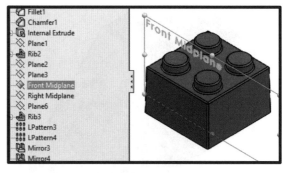

The line on the face is collinear with the face, not vertical. If after you create the revolve, the feature tree shows two solid bodies, then edit your sketch and make the line on the face collinear with edge of the fillet.

Figures 4.92 and 4.93

Chapter 4: LEGO Minions, Part 1—Designing **103**

➤ Add a 0.1 chamfer feature to the three locations as shown in Figure 4.94.

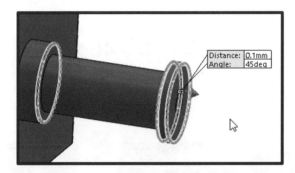

Figure 4.94

➤ Add a 0.1 fillet feature to the three locations as shown in Figure 4.95.

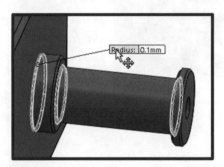

Figure 4.95

➤ Select the face of the revolve and create an **Extruded Cut** feature.
➤ Sketch a **Center Rectangle** on this face.
➤ Add the relationships and dimensions as shown in Figure 4.96 to fully define the sketch.

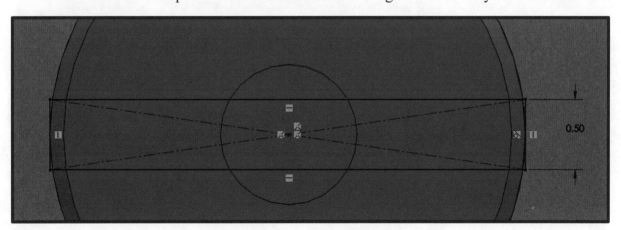

Figure 4.96

➤ Exit the sketch and make the extrusion depth 3.0.

Chapter 4: Lego Minions, Part 1—Designing

Full Round Fillet

➢ Select the **Fillet** feature.

➢ Under the **Fillet Type**, select the **Full round fillet** ⦿ Full round fillet radio button. *For SolidWorks 2015:* select the icon.

➢ Select the faces of the part as shown in Figure 4.97 and accept these selections.

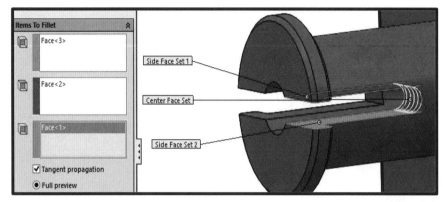

It does not matter in which order you select the face sets, as long as the second face selected is in the middle of the first two.

You can right click after selecting a face to end the selection.

Figure 4.97

To see this style of fillet being created, watch the video "**Create Full Round Fillet**."

➢ Repeat this fillet for the other side of the slot.

➢ **Save your work.**

5: Design Tables with Excel

Standard

Design tables can make the creation of similar parts a quick task. And these can be easily created and modified in an Excel spreadsheet.

Table 5.1 shows the sizes of standard center drills, such as the model of one without flutes as shown in Figure 5.1. In order to make the creation of the design table simpler, each feature that will be changed as the drill size changes will be named. Table 5.1 shows the sizes of bell type center drills.

Size	Body Dia (A)	Drill Dia (B)	Drill Length (C)	Overall Length (D)
00	1/8	0.025	0.030	1 1/8
0	1/8	1/32	0.038	1 1/8
1	1/8	3/64	3/64	1 1/4
2	3/16	5/64	5/64	1 7/8
3	1/4	7/64	7/64	2
4	5/16	1/8	1/8	2 1/8
5	7/16	3/16	3/16	2 3/4
6	1/2	7/32	7/32	3
7	5/8	1/4	1/4	3 1/4
8	3/4	5/16	5/16	3 1/2

Table 5.1

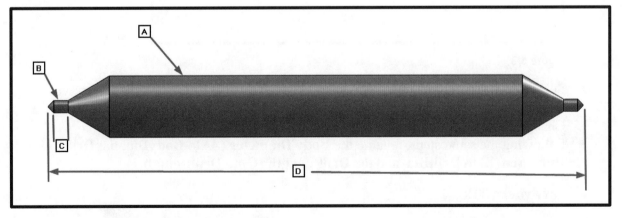

Figure 5.1: Standard Center Drill (without flutes)

> Start a new inch drawing.

Chapter 5: Design Tables with Excel

➢ Create a revolved part to represent the 00 size center drill as shown in Figure 5.2.

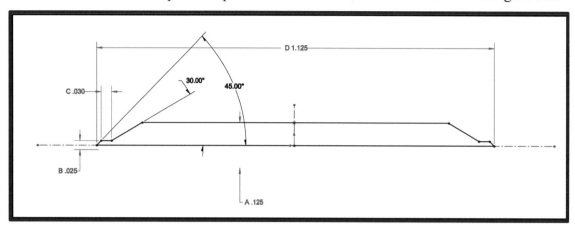

Figure 5.2

If you are not sure how to do this watch the video "**CD 00 Creation**."

➢ Double click on the part, you should see all of the dimensions appear as in Figure 5.3.

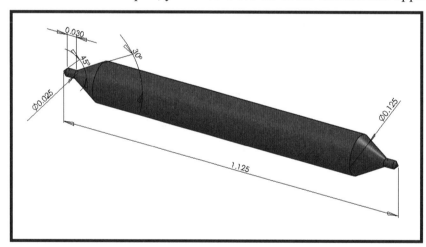

Figure 5.3

➢ Select the 1.125 overall length dimension. On the left-hand side of your screen, you should see the heading **Primary Value (D)** and a value similar to: **D5@Sketch3**. Change the **D5** to **OAL**.

➢ Following the same steps, rename the **Body Diameter (A)** to **BodyDia**, the **Drill Diameter (B)** to **DrillDia** and the **Drill Length (C)** to **DrillLength**.

➢ **Save your work.**

To see this being done, watch the video "**CD Naming Dimensions**."

Creating the Design Configurations

➢ From the drop down menu, select **Insert**, **Tables, Design Table**

➢ Accept the default table creation styles
➢ In the **Dimensions** window, select the four dimensions that you named as the items to be included in the table and select **OK**

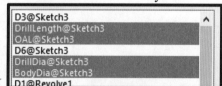

*You may need to resize the Excel window that opens up. **If you do, be very careful.** If you accidentally click outside of the window, the design table will close, and you will lose all of your work and have to start again.*

To resize the window carefully select the lower right-hand corner, click on it, and hold to drag and make it larger. Release the mouse when you are satisfied with your new size. If entering the fractional values into the table, use an equals sign or Excel may recognize these figures as dates rather than as numerical values.

➢ Set up the table as shown in Figure 5.4. *Note that the columns are not in the same order as shown in Table 5.1.*

	A	B	C	D	E
1	Design Table for: Center Drill				
2		DrillLength@Sketch3	OAL@Sketch3	DrillDia@Sketch3	BodyDia@Sketch3
3	Default	0.0300	1.1250	0.0250	0.1250
4	0	0.0380	1.1250	0.0313	0.1250
5	1	0.0469	1.2500	0.0469	0.1250
6	2	0.0781	1.8750	0.0781	0.1875
7	3	0.1094	2.0000	0.1094	0.2500
8	4	0.1250	2.1250	0.1250	0.3125
9	5	0.1875	2.7500	0.1875	0.4375
10	6	0.2188	3.0000	0.2188	0.5000
11	7	0.2500	3.2500	0.2500	0.6250
12	8	0.3125	3.5000	0.3125	0.7500

Figure 5.4

To see this being done, watch the video "**CD Design Table**."

Chapter 5: Design Tables with Excel

> Click in the graphics area outside of the table to finish creating your design configurations.
> You should now see a message as shown in Figure 5.5, select **OK**.

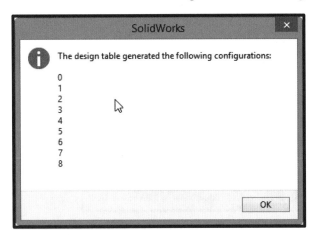

Figure 5.5

> Your design configurations should look similar to Figure 5.6, with the "**X**" indicating they were created in Excel.

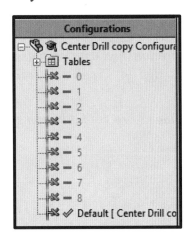

Figure 5.6

> Double click on each design configuration to ensure that each part is correct and there are no rebuild or size errors. If there are no rebuild errors, then each configuration will have a check mark beside it

*If this step fails, start again. However, if the **Design Table***  *option is grayed out, then a partial table was created.*

*If you have errors, expand the **Tables** folder* 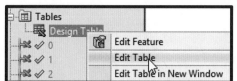 *and right click on the **Design Table** and select **Edit Table**. Or select **Edit Table in New window** to open an Excel spreadsheet to work in*

> Save your work.

Bell Type

Center drills from size 1 and up are available in what is known as the bell type.

Size	Body Dia (A)	Drill Dia (B)	Drill Length (C)	Overall Length (D)	Bell Dia (E)
11	1/8	3/64	3/64	1 1/4	0.1
12	3/16	5/64	5/64	1 7/8	0.15
13	1/4	7/64	7/64	2	0.2
14	5/16	1/8	1/8	2 1/8	0.25
15	7/16	3/16	3/16	2 3/4	0.35
16	1/2	7/32	7/32	3	0.4
17	5/8	1/4	1/4	3 1/4	0.5
18	3/4	5/16	5/16	3 1/2	0.6

Table 5.2

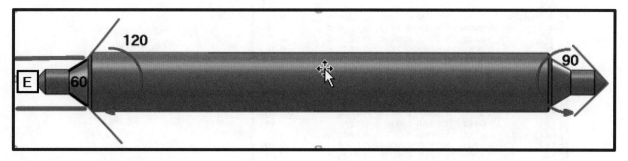

Figure 5.7: Bell Type Center Drill

110 Chapter 5: Design Tables with Excel

Adding a Feature to the Table

> Add the bell feature to your number one center drill configuration. (*Hint: it is another revolved feature.*)
> Name this feature **Bell**.
> To select the number 1 configuration to work on, double click on the **1** in the **Configuration Manager** tree:

If you are not sure how to do this watch the video "**CD Bell**."

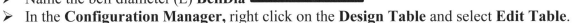

> Name the bell diameter (E) **BellDia**
> In the **Configuration Manager,** right click on the **Design Table** and select **Edit Table**.
> When the window to **Add Rows/Columns** opens, select **BellDia@Sketch#** in the **Parameters** section and click **OK**.
> Add the sizes 11–18 to the chart.
> Two columns will have to be added to the design table, so that it looks like Figure 5.8.

	A	B	C	D	E	F	G
1	Design Table for: Center Drill						
2		DrillLength@Sketch3	OAL@Sketch3	DrillDia@Sketch3	BodyDia@Sketch3	$STATE@Bell	BellDia@Sketch6
3	Default	0.0300	1.1250	0.0250	0.1250	S	0.10
4	0	0.0380	1.1250	0.0313	0.1250	S	0.10
5	1	0.0469	1.2500	0.0469	0.1250	S	0.10
6	2	0.0781	1.8750	0.0781	0.1875	S	0.15
7	3	0.1094	2.0000	0.1094	0.2500	S	0.20
8	4	0.1250	2.1250	0.1250	0.3125	S	0.25
9	5	0.1875	2.7500	0.1875	0.4375	S	0.35
10	6	0.2188	3.0000	0.2188	0.5000	S	0.40
11	7	0.2500	3.2500	0.2500	0.6250	S	0.50
12	8	0.3125	3.5000	0.3125	0.7500	S	0.60
13	11	0.0469	1.2500	0.0469	0.1250	U	0.10
14	12	0.0781	1.8750	0.0781	0.1875	U	0.15
15	13	0.1094	2.0000	0.1094	0.2500	U	0.20
16	14	0.1250	2.1250	0.1250	0.3125	U	0.25
17	15	0.1875	2.7500	0.1875	0.4375	U	0.35
18	16	0.2188	3.0000	0.2188	0.5000	U	0.40
19	17	0.2500	3.2500	0.2500	0.6250	U	0.50
20	18	0.3125	3.5000	0.3125	0.7500	U	0.60

Figure 5.8

Chapter 5: Design Tables with Excel **111**

- To add the column you can type in the cell, **exactly** the feature name, such as in cell **F2**, type: **$STATE@Bell**. The **S** tells SolidWorks that this feature will be suppressed in this configuration, the **U** tells SolidWorks that this feature will be unsuppressed in this configuration; therefore, in the unsuppressed configuration, the bell will be seen.

    *Alternatively, to add the **$STATE@Bell** to the chart, select the cell in the chart and then double click on the bell feature in the graphics area.*

- Click in the graphics area to exit the Excel Window.
- Select **OK** when SolidWorks tells you that configurations were generated:

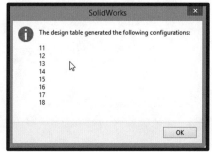

- Verify that each configuration has been updated correctly.

- **Save your work.**

To see this being done, watch the video "**CD Edit Table**."

6: Heat Sink

Creating Ribs

> Start a new metric part.
> In the **Top plane,** create an extrusion 46.8 x 44.6 x 1.8, as shown in Figure 6.1 (use

Center Rectangle)

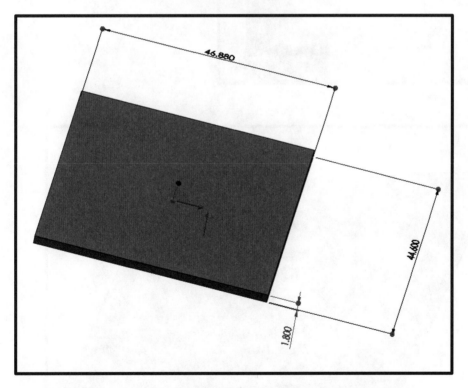

Figure 6.1

> **Save your work** as **YouName_Heat Sink**.

113

Chapter 6: Heat Sink

➢ Create a plane offset **1.4 mm** from the center of the **46.88 mm** side of your part, as shown in Figure 6.2 below.

   o Select the **Right** plane in the feature tree

   o Select **Reference Geometry, Plane**
   o Enter **1.4** as the offset distance and accept this:

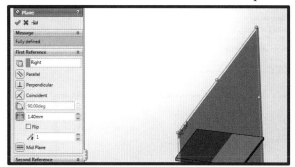

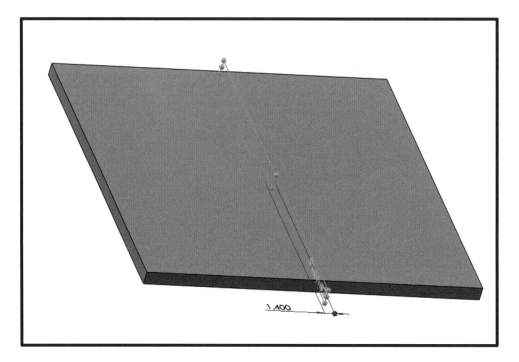

Figure 6.2

Chapter 6: Heat Sink 115

➢ Create a sketch in this new plane, as shown in Figure 6.3. Make certain that the sketch is fully defined.

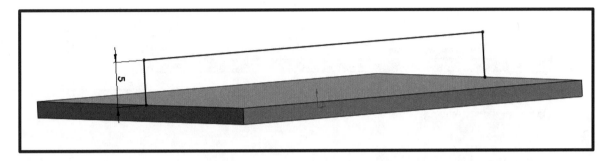

Figure 6.3

o Right click on the plane and select **Sketch**

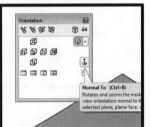

o Press the **Space Bar** and select **Normal To**

➢ Select this sketch from the feature tree

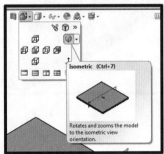

➢ Switch to the **Isometric** view

116 Chapter 6: Heat Sink

➢ Select the **Rib** feature from the **Features** ribbon bar as shown in Figure 6.4.

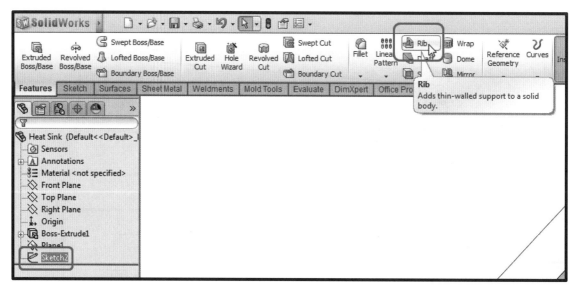

Figure 6.4

➢ Set up the rib so that it is offset on both sides of your sketch, is **0.8 mm** wide, and will be created towards the part and accept these selections, as shown in Figure 6.5.

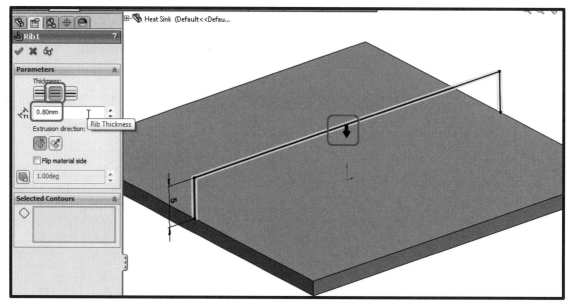

Figure 6.5

➢ Your part should look similar to Figure 6.6.

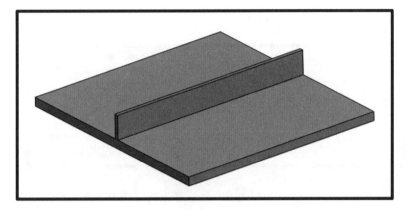

Figure 6.6

To see this being done, watch the video "**Heat Sink Rib**."

➢ Add a **0.5 mm** fillet to each side of the rib, so that your part now looks similar to Figure 6.7.

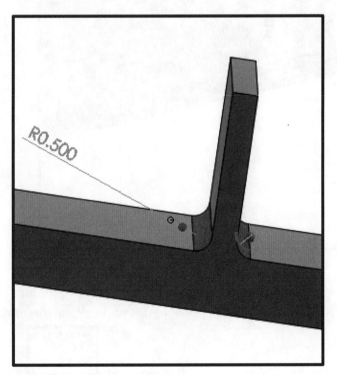

Figure 6.7

Chapter 6: Heat Sink

You will now create a linear pattern of this rib.

> From the **Feature** tab select **Linear Pattern.**

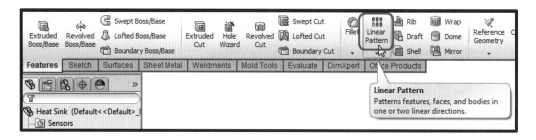

> Set up the **LPattern** properties as shown in Figure 6.8 and accept these selections.

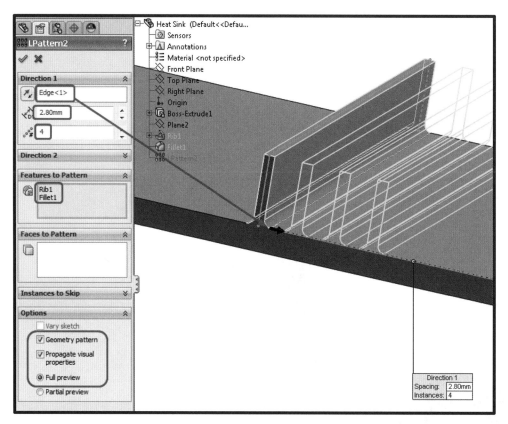

Figure 6.8

When selecting the direction vector, be certain to select the part edge near where the pattern will be created from, as this will give it the proper sense of direction. If the direction is wrong, it can be flipped by clicking on the arrow or by clicking on the direction icon

To see this being done, watch the video "Rib Linear Pattern."

Now you will add a full round radius to the top of the initial rib and then add it to the linear pattern.

➢ Move your mouse over the **Rollback Bar** at the bottom of the Feature Manager Design Tree as shown in Figure 6.9. Notice how the mouse pointer changes to the shape of a hand pinching the bar.

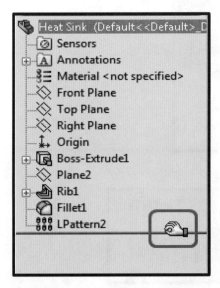

Figure 6.9

➢ Drag the bar so that it is above the "LPattern" and below the "Fillet" feature. To drag the bar, click and hold the mouse.

To see this being done, watch the video "**Rollback Bar.**"

➢ Create a Full Round Fillet at the top of the rib.

- o Select **Fillet**
- o Select **Full round fillet**

To see this being done, watch the video "**Create Full Round Fillet**."

120 Chapter 6: Heat Sink

- Select the face as shown in Figure 6.10, as **Face Set 1.**

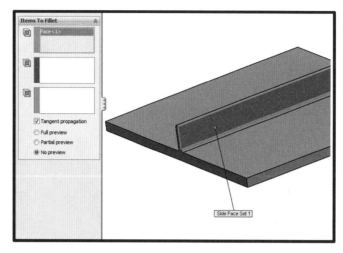

Figure 6.10

- Click in the **Center Face Set** selection area
- Select the top face as shown in Figure 6.11.

Figure 6.11

- Click in the **Face Set 2** area

o Select the side opposite the first face as the **Side Face Set 2** as shown in Figure 6.12 and accept ☑ these selections.

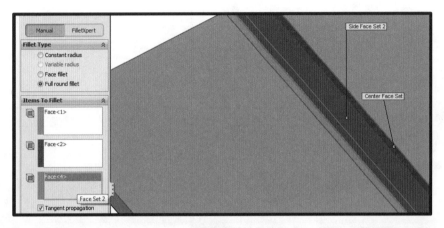

Figure 6.12

➢ **Save your work.**

➢ Move the **Rollback Bar** to the bottom of the feature list and edit (right click on the LPattern feature and select **Edit Feature**) the Linear Pattern so that it now includes the full round fillet.

If you are not certain how to do this watch the video "**Edit LPattern**."

Your part should now look similar to the Figure 6.13.

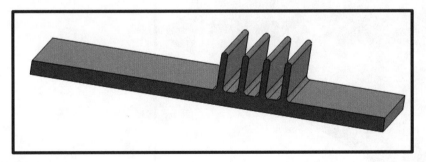

Figure 6.13

- Create a new plane **15.5 mm** from the Right Plane as shown in Figure 6.14.

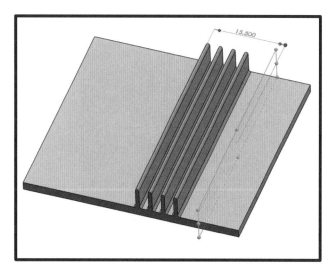

Figure 6.14

If you cannot remember how to do this watch the video "**15_5 Plane**."

- Create another work plane that is offset **3 mm** from this last plane as shown in Figure 6.15.

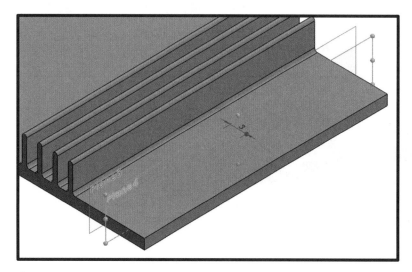

Figure 6.15

- In this last plane that you created, create another rib using the Convert Entities command.

If you are not certain how to do this, watch the video "**Rib Convert Entities**."

Chapter 6: Heat Sink   **123**

➢ **Save your work.**

➢ Edit this last rib to look like the Figure 6.16, try to do this using one Extrusion and one **Extruded Cut** feature. (The radius is tangent to the top and side of the rib.)

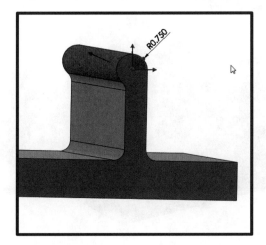

Figure 6.16

You will now mirror this last rib through the plane located **15.5 mm** from the part center.

➢ Select the Mirror command:

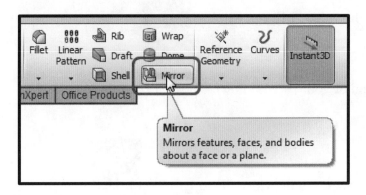

124 Chapter 6: Heat Sink

➢ Set up Mirror options as shown in Figure 6.17 and accept these selections. The mirror plane is **15.5 mm** from the right plane. The Features to Mirror are all of the features that make up your last rib.

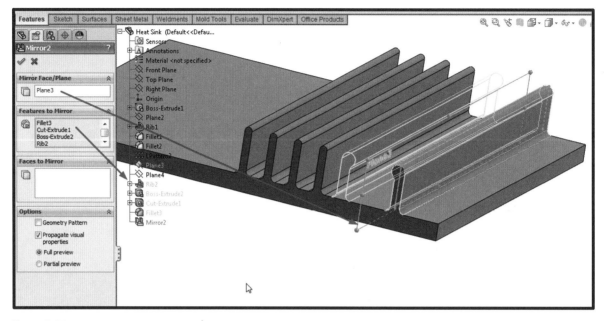

Figure 6.17

To see this being done, watch the video "**Mirror Rib**."

Create two more ribs on the end of your part as shown below. (The lower rib should extend **2.5 mm**.)

If you are not sure how to do this, watch the video "**End Rib**."

➢ **Save your work.**

> Your part should now look similar to Figure 6.18.

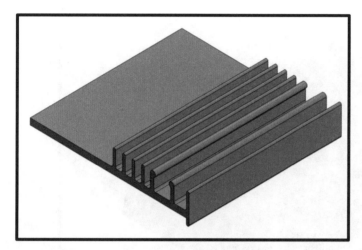

Figure 6.18

> Use the above techniques to make copies of all of the ribs to the others side of the part. Your final part should look similar to Figure 6.19.

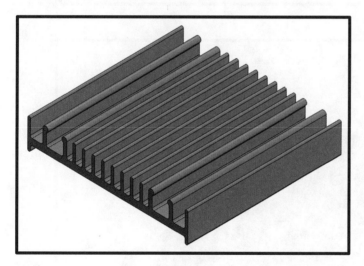

Figure 6.19

> **Save your work.**

Extra Practice

Create the following part, as shown in Figures 6.20 and 6.21 on the next page, using one extrusion, one shell, one rib, and one linear pattern.

Chapter 6: Heat Sink

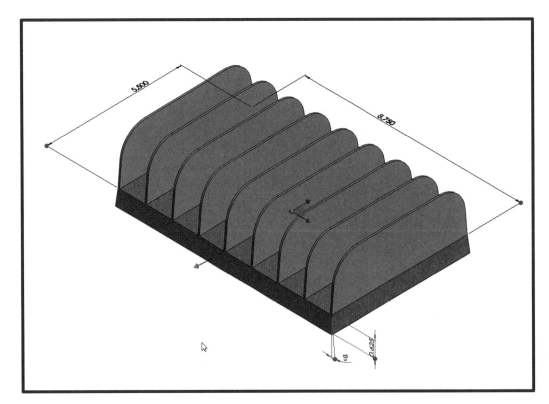

Figure 6.20

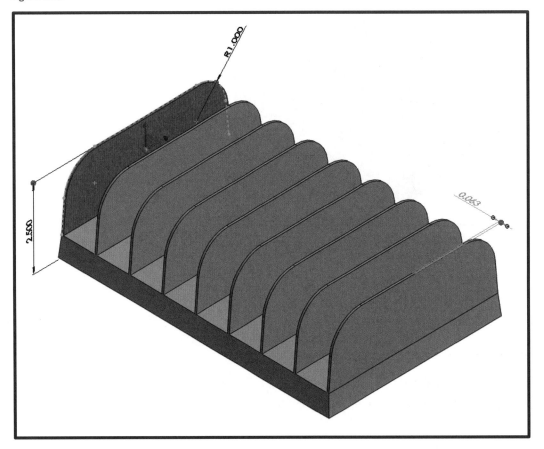

Figure 6.21

7: LEGO Minions, Part 2—Assemblies

Assemblies

An assembly can be started by choosing to start a new blank assembly or by selecting to create an assembly from an open part. Assemblies may contain only a few or numerous parts and may have multiple subassemblies. They can be modeled to behave as the actual functioning part would, so that motion studies can be performed on the assembly. This also enables you to create professional assembly instructions and videos of how the parts should come together. Unlike a part drawing, assemblies may or may not need to be fully defined.

> Open your **BlockForWheel**.

> Select **File, Make Assembly from Part**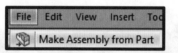

*On the left side of your screen, you will see the **Begin Assembly** property manager. Here a list of all open parts and/or assemblies will be displayed in the **Part/Assembly to Insert** section. The part you launched the assembly creation from will be highlighted by default.*

> **In SolidWorks 2014:** select **OK** to insert this part into the assembly.

> **In SolidWorks 2015:** you have the option to change the orientation of the part by rotating it about the X,Y, or Z axis prior to selecting **OK:**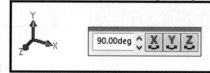

In the feature tree, you should notice that your part name has the letter f in front of it , *this means that the part is fixed and will not move. This part's origin will be fixed to the assemblies' origin, and its front plane will be fixed to the assemblies' front plane, as will its right and top planes be fixed to the assemblies' right and top planes.*

*If you do not wish this part to be fixed, right click on it and select **Float***

> Save this as **WheelEye_YourName**.

128   Chapter 7: LEGO Minions, Part 2—Assemblies

Inserting a Component

> From the **Assembly** tab select **Insert Components**
> Select **Browse** (this can be found at approximately the middle of the **Insert Components** property manager).

If the next component that you wish to add to the assembly is already open, then it will appear in the list of the Part/Assembly to Insert section and can be selected there.

> Navigate to your **Hub** and select it.
> Click in the graphics area near the block to place the hub in the assembly.

Your screen should now look similar to Figure 7.1.

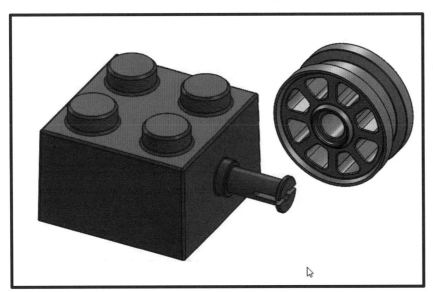

You can click on and drag this part anywhere you like in the assembly, because it is not yet matted.

Figure 7.1

> Select the **Hub** in the feature tree.

> Select the drop down arrow below **Move Component** 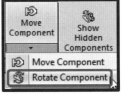 and then select **Rotate Component.**

> In the **Rotate Component** options select **Free Drag**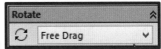

➢ Click and hold the mouse button near the component and move the mouse until

the part is in the approximate orientation as shown in Figure 7.2 and accept this

Figure 7.2

Do not worry about getting the part positioned exactly where you would like it. Shortly, you will be adding mates, which will fully define your parts.

To see this being done, watch the video "**Rotate Assembly Component**."

Adding Mates

Mates define how assembly components can behave in the assembly. We add them in an attempt to make our model behave exactly as the real assembly.

Concentric

➢ Select the round face of the block as shown in Figure 7.3 and then select the **Mate** option.

Figure 7.3

*Like most SolidWorks commands, you can also start a mate without preselecting geometry and simply selecting the **Mate** option. By preselecting the geometry, it will appear in the mate property manager.*

➢ Select any cylindrical face on the hub as shown in Figure 7.4.

When you select two cylindrical faces, SolidWorks will assume you want to add a **Concentric** mate and highlight it for you.

If you accept this mate, then these two part surfaces will now share a center. If you do not wish the part to rotate then select the **Lock Rotation** option.

Figure 7.4

➢ Accept the **Concentric** mate and exit the mate property manager.

To see this being done, watch the video "**Adding a Concentric Mate**."

Expand both parts' feature trees. Just below each part's name, you should notice a folder called **Mates in WheelEye**. In this folder, all mates used to define this part can be found. At the bottom of the feature tree, you will find a **Mates** section containing all of the assemblies' mates. Any mate can be suppressed by clicking on it and selecting **Suppress**. Select the mate from one of these locations and then select **Edit Feature**. In the **Concentric** property manager, you will see both of the selected features used in the mate; these can be found in the **Mate Selections** area. Right click on one of the mate selections, and you will be given the option to **Delete** that selection and then replace it. Or you can select **Clear Selections** and remove all selected entities, allowing you to reselect all mate entities.

Chapter 7: LEGO Minions, Part 2—Assemblies   **131**

- ➢ Exit ❌ the mate property manager.
- ➢ With your cursor on the hub, click and hold down the mouse button. Move your mouse around. You should notice that the part can move away from the block or into the block.

Editing a Part in the Assembly

- ➢ In the feature tree, select your **BlockForWheel** part,

- ➢ Select **Edit Part**. *(You should notice the hub becomes transparent and in the feature tree all of the hub's features are blue.)*

- ➢ Add a mid-plane as shown in Figure 7.5.

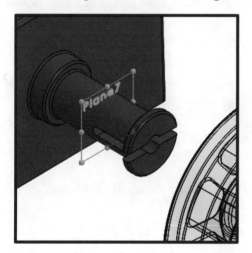

Figure 7.5

- ➢ Select **Edit Component** from the upper left-hand corner or from the upper right-hand corner of the graphics area to stop editing this component and return to the assembly.

To see this being, done watch the video "**Editing a Part in an Assembly**."

132   Chapter 7: LEGO Minions, Part 2—Assemblies

Coincident

> ➢ In the feature tree, expand the **BlockForWheel** section, select your newly created plane,

and then select **Mate**

> ➢ Expand the feature tree ![WheelEye], which is now found on the left side of the graphics area.
> ➢ Scroll down the feature tree by rolling the mouse wheel while the mouse is in the feature tree area of the graphics screen. Select the **Front Plane** of the hub as shown in Figure 7.6.

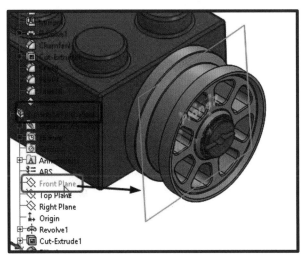

Alternatively, you can turn on the visibility of the hub's front plane before starting the mate and then select the plane in the graphics area.

After adding this mate, try to move the hub. It should only rotate about the axis, as the actual part would.

Figure 7.6

> ➢ Accept the **Coincident** mate [Coincident] and exit [X] the mate property manager.

To see this being done, watch the video "**Coincident Mate**."

> ➢ **Save your work.**

At any point, if you wish to collapse all items in the feature tree, right click and select **Collapse Items** [Collapse Items]. *Conversely, if you wish to expand all items, select the part or assembly name* [WheelEye (Default<Display State-1>)] *at the top of the feature tree and then press the asterisk (*, shift and 8) key on your keyboard.*

➢ Insert your wheel into the assembly and add a **Concentric** mate to the hub. This time, enable the **Lock Rotation** option. (*This will stop it from rotating on the hub but will not stop the hub's rotation.*)

➢ Add a **Coincident** mate between the top plane of the wheel and the front plane of the hub. Your assembly should now look similar to Figure 7.7.

This will become a subassembly once it is inserted into another assembly.

Figure 7.7

➢ **Save your work.**

Creating a Subassembly within an Assembly

At this point, you may realize that you meant to make the wheel and the hub a separate subassembly. If so, this can easily be done now. Sometimes, when we create a subassembly after the fact, the mates will require reworking.

➢ Hold the **CTRL** or the **SHIFT** key and select your **Hub** and **Wheel** in the feature tree.

➢ Right click and select **Form New Subassembly Here**. (You may have to select the down arrow before you can see this option.)

➢ Rename this new subassembly **WheelAndHub**

To see this being done, watch the video "<u>Creating a Subassembly within an Assembly</u>."

134   Chapter 7: LEGO Minions, Part 2—Assemblies

➢ **Save your work**: A message will open up asking you how you wish to save your work, as shown in Figure 7.8. Select **Save externally** and specify where to save the subassembly. *(By default, SolidWorks will point to where your parts are saved, this is the best location to save your parts.)*

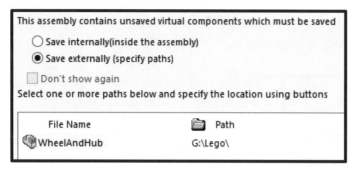

If you select **Save internally**, then you will not be able to create drawings of this subassembly. However, you can always save it externally later.

Figure 7.8

➢ Close this assembly.

Creating your Minions

➢ Select **File, New**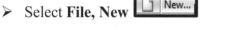

➢ Select **Assembly** as the **New SolidWorks Document** type
➢ Browse to your **TallBlock** and place it in the assembly as the base component.
➢ Select the **TallBlock** in the feature tree.
➢ A new drop down menu will appear, allowing you to select which configuration you would like.
➢ Select the **2x4** configuration and accept this selection
➢ Right click on the part name in the feature tree and select **Float**
➢ Rotate your part so that in the isometric view it looks similar to Figure 7.9.

Figure 7.9

Chapter 7: LEGO Minions, Part 2—Assemblies

- Add three **Coincident** mates as follows:
  - Right assembly plane to the block's front plane
  - Top assembly plane to the block's top plane
  - Front assembly plane to the block's right plane
- Exit the mate property manager.
- **Save this** as **MyMinion.**

Copying an Existing Assembly Component

- While holding the **CTRL** key on your keyboard, select the **TallBlock** part, drag it into the graphics area, and release the mouse to place a copy of this into the assembly.

To see this being done, watch the video "**Copying an Assembly Component**."

- Select the **Default** configuration for this part.

Mating Surfaces/Select Other

- Select the top face of the **2x4** block and select the **Mate** icon as shown in Figure 7.10.

As your assemblies become larger, it is cumbersome to rotate the entire assembly in order to be able to select a hidden face. SolidWorks has several features to make the selection of hidden features quick and easy. You can easily set an object's visibility to transparent or hidden, and you can use the Select Other option.

Figure 7.10

- Look **Normal To** the selected face.
- Right click on the top face of the **Default** block and choose **Select Other**. *(The top face will now become see-through, allowing you to select unseen geometry.)*

136   Chapter 7: Lego Minions, Part 2—Assemblies

➢ Select the bottom face of this part as shown in Figure 7.11 and add the **Coincident** mate.

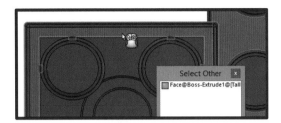

Figure 7.11

To see this being done, watch the video "<u>Select Other</u>."

➢ Add two more **Coincident** mates between these two parts, mating each part's **Right Midplane** to the other part's **Midplane** and each part's **Top Midplane** to the other part's **Top Midplane.**

Copy with Mates

➢ Right click on the **Default** block in the feature tree.
➢ Select **Copy with Mates**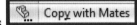
➢ *In SolidWorks 2015:* The window will show all items selected to be copied. Press the right arrow to go to the screen where the mates can be applied.
➢ The three mates involved in matting the original two parts will appear in the **Copy with Mates** property manager. For each mate, you can choose to repeat it, to not use it, or to select a new mate entity.

  o For the last two mates, check the box to **Repeat** them
  o For the first mate, select the top face of the original **Default** block as shown in Figure 7.12, accept these selections, and exit the copy with mates function. Your part should now look similar to Figure 7.13.

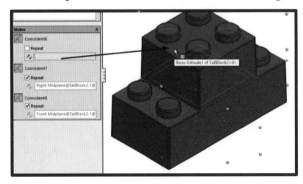

Figure 7.12

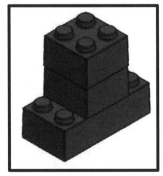

Figure 7.13

Chapter 7: LEGO Minions, Part 2—Assemblies   **137**

To see this being done, watch the video "<u>Copy with Mates</u>."

Configuring a Part Color Only for the Assembly

> ➤ Right click on the **2x4** tall block, select the drop down arrow beside the appearance icon (the beach ball), and then select the **TallBlock** with the assembly icon beside it as shown in Figure 7.14.

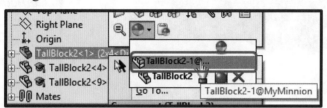

The part appearance can be configured different in the assembly than in the part when desired.

Figure 7.14

> ➤ Select the **Standard ColorBlack**  and accept this selection.
> ➤ Select both of the **Default** blocks in the feature tree and change their color to blue.

> ➤ On the right side of your screen, select the **Appearances, Scenes, and Decals** tab.
> ➤ Expand the **Lights** section and select **LED**.

> ➤ Select the **Blue LED** 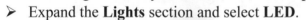 appearance and accept these selections. *(This will increase the brightness of your part.)*

Adding a Part through File Explorer

> ➤ From the right side of your screen, select the **File Explorer** tab:

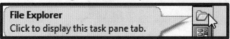

> ➤ Navigate to the folder where your parts are saved. Notice that if you hover over a part the

picture of the part will appear

> ➤ Drag and drop your short block into the assembly.
> ➤ From the **Select a Configuration** window, choose your **4x6** configuration and **OK**.

➤ Add two **Coincident** mates to the short block as shown in Figure 7.15. *(Depending on how you created your short block you may need to create a new plane in this part.)*

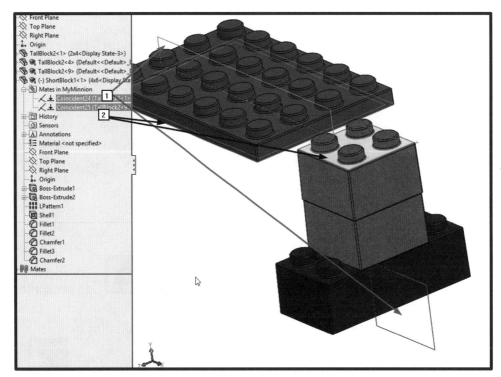

Coincident mate 1 is between two planes. Coincident mate 2 is between the faces of the blocks.

Figure 7.15

Tangent Mate

➤ Select the inside face of the short block and select the **Mate** option as shown in Figure 7.16.
➤ Select the round face of the default tall block as shown in Figure 7.17.

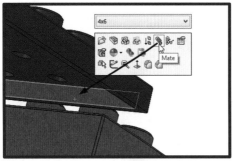

Figure 7.16

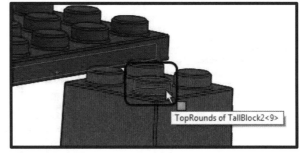

Figure 7.17

Chapter 7: LEGO Minions, Part 2—Assemblies

➢ Add the **Tangent** mate as shown in Figure 7.18 and exit ![x] the mate property manager.

Figure 7.18

➢ **Save your work.**

Copying an Appearance

➢ Click on one of the blue tall blocks and select **Copy Appearance**

➢ Click on the short block and select **Paste Appearance**

Mirror Assembly Components

➢ Select (or create) the plane in the middle of your short block as shown in Figure 7.19.

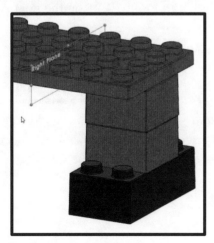

Figure 7.19

➢ Select the drop down arrow below **Linear Component Pattern** and select **Mirror Components** 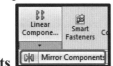 . *(This will establish the mirror plane.)*

➢ Select the first three blocks as the **Components to Mirror**
➢ Accept these selections, and your work should look similar to Figure 7.20.

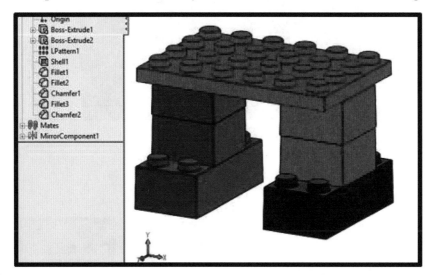

Figure 7.20

To see this being done, watch the video "**Mirror Assembly Components**."

➢ Apply the appropriate appearance to each of the copied parts.
➢ Add another **4x6** block, fully mate it in place, as shown in Figure 7.21 on the next page and change its appearance to blue.

*After mating this block, click on it and select **Change Transparency*** *. This will allow you to see through the now transparent block, and you can visually check for any interference of parts, as shown in Figure 7.22 on the next page. When you are ready, click on the part again, either in the graphics area or in the feature tree, and turn off the **Change Transparency**.*

Chapter 7: LEGO Minions, Part 2—Assemblies

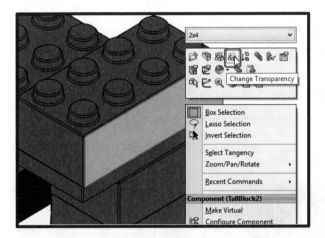

Figure 7.21

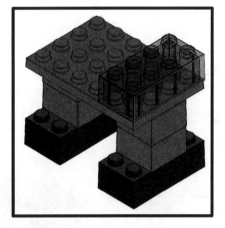

Figure 7.22

Assembly Linear Pattern

Evaluate/Measure

- Select the **Evaluate** tab
- Select the **Measure** tool.
- Select one edge to measure its length or select two edges as shown in Figure 7.23 to know the distance between two features. Close the measure window to exit the command.

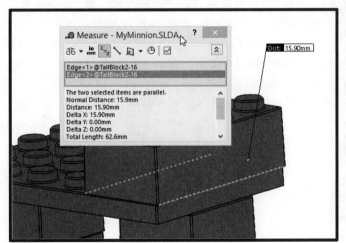

Figure 7.23

With the measure tool, you can measure angles, lengths, distances and much more.

To see this being done, watch the video "**Evaluate Measure**."

➢ Select **Linear Component Pattern**
➢ Pattern the last block that you added so that your part looks similar to Figure 7.24. The pattern distance is **15.9 mm**, and there should be three instances of this component.

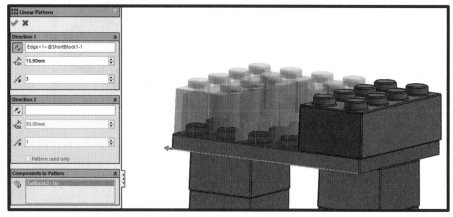

The only difference between this pattern feature and the part feature pattern is that, in an assembly, you can only choose to pattern components.

Figure 7.24

➢ Apply the blue color to these blocks.
➢ Insert a **OneBy** block into the assembly.
➢ Select the **1x2** configuration.
➢ Change the appearance to blue and fully mate it in place, using two **Coincident** and one **Tangent** mate. Before applying the **Tangent** mate, move the block so that the center is past the edge of the round feature as shown in Figure 7.25.

*The tangency is to the interior wall of the block. By placing the block on the correct side of the round feature, the **Tangent** mate should be correctly applied, because the **Tangent** mate will mate to the nearest side of the round part.*

When adding mates, you may notice the mate is the opposite of what you want. When this is the case, you can flip the mate alignment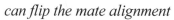

Figure 7.25

➢ Select the **LocalLPattern1** in the feature tree and then select **Edit Feature**:

➢ Add the **1x2** block to the pattern.

- Mirror the three **1x2** blocks so that your assembly now looks similar to Figure 7.26.
- Change the color of these blocks to blue as shown in Figure 7.27.

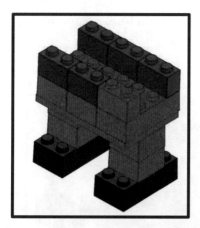

Figure 7.26

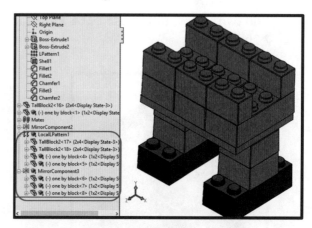

Figure 7.27

- **Save your work.**

*You can expand the pattern and mirror assembly features: hold the **CTRL** or the **SHIFT** key and then select multiple parts to apply the appearance. The copy and paste appearance option will save time when altering appearances.*

Show/Hide a Part

As your assembly becomes larger, you may find some parts are making it difficult to see.

- Select the first **1x2** block, either from the feature tree or from the graphics area.
- Select **Hide Components** 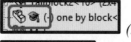, and this block will disappear from your assembly.
- To turn the visibility of the block back on, select it from the feature tree  *(notice the appearance)* and then select **Show Components**:

*In larger assemblies, you may forget what and how many components are hidden. To see them, select **Show Hidden Components** from the Assembly ribbon bar.*

- Add two more fully mated **2x4 TallBlocks.** Note that you can add one, mate it, and then edit your initial mirror to include this block.

➢ Set the color of these two blocks to **Yellow LED**. Your assembly should look similar to Figure 7.28.

➢ Add two, black, fully mated **2x4 TallBlocks**. Your part should look similar to Figure 7.29.

Figure 7.28

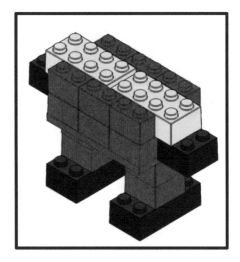

Figure 7.29

➢ Add the blocks to the next layer as shown in Figure 7.30, set the colors as shown, and all blocks should be fully mated. *This means they are unable to move if you select them and try to drag the part around*

You should recall that, when a part sketch was not fully defined, it had a minus (–) sign beside the sketch in the feature tree. The same is true for a part in an assembly. When it is not mated so that it cannot move, there will be a minus sign beside the part's name in the feature tree

Figure 7.30

You may also wish to view the temporary axes and use them for mate creation. Depending on how you have mated entities, you may have accidentally forced blocks to physically interfere with one another. This is not possible with the actual blocks.

Interference Detection

- From the **Evaluate** tab select **Interference Detection**
- Select **Calculate**
- A list of all of the interfering components and the interference amounts will appear as shown in Figure 7.31.

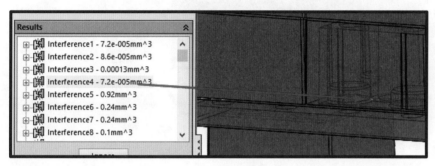

When an interference is selected, the two components involved will become transparent, and the location of the interference will show as red with a pink outline.

Figure 7.31

- The interference shown in Figure 7.31 can be ignored for a plastic part, as it will not stop the parts from functioning. Select the **Ignore** option for all interferences that are less than **0.001 mm**. *(Note: Larger interferences will need to be corrected.)*
- Larger interferences, such as the one shown in Figure 7.32, are often due to mating errors, and can easily be fixed by altering the mates.

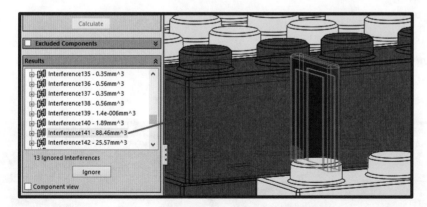

Figure 7.32

146 Chapter 7: LEGO Minions, Part 2—Assemblies

➢ Other mates, such as the one shown in Figure 7.33, reveal that we have a design problem. We need to repair this design issue and continue on, happy with the knowledge that the issue was caught before the actual components were produced.

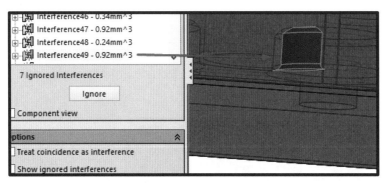

Figure 7.33

Open Part

➢ From the feature tree or from the graphics area, select one of your tall blocks and then select **Open Part**

➢ Change the diameters of the **Internal Extrude** feature from **6.51** to **5.90** and from **4.80** to **4.30**.
➢ Rebuild the part and save the changes.
➢ Close the part.
➢ Rebuild your assembly.

Parallel Mate

➢ For your first **OneBy 1x2** block, added to this last level, edit the mates so that you have the following mates:
  o **Coincident** between the two surfaces as shown in Figure 7.34

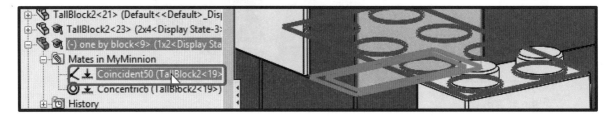

Figure 7.34

Chapter 7: LEGO Minions, Part 2—Assemblies **147**

- o **Concentric** as shown in Figure 7.35. *(The mate could also be between the other two round bosses.)*
- o Add a **Parallel** [Parallel] mate between the two part edges as shown in Figure 7.36.

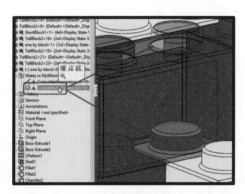

Figure 7.35

Figure 7.36

Do not worry about further interference issues for this assembly.

➤ Add a yellow **4x9 Short Block** that is fully mated to the top of your assembly and is coincident with the previous 4x9 short block.
➤ Add and fully mate another layer with the following blocks, in the orientation as shown in Figure 7.37:
  - o Two yellow **2x4 Tall** blocks
  - o Two yellow **1x2 Tall** blocks
  - o One yellow **1x2 Tall Angled** block
  - o One red **1x2 Tall** block

Figure 7.37

➤ Add another layer of yellow, fully mated tall blocks as shown in Figure 7.38.

Figure 7.38

Inserting a Subassembly

- Select **Insert Components**
- **Browse** to where your parts are saved.
- In the **Open** part window, turn on the **Quick Filter** for **Assemblies** only
- Select your **WheelEye** subassembly and click in the graphics area to place it.
- Fully mate this subassembly as shown in Figure 7.39.

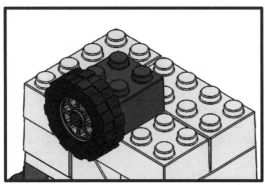

*Try to rotate the wheel. It should not move. This is due to the fact that SolidWorks will make a subassembly **Rigid** when you bring it into an assembly, making it easier to mate.*

Once the subassembly is made flexible, it can then again behave as it was designed to. In this case, the wheel can rotate.

Figure 7.39

Flexible Subassembly

- In the feature tree, click on your **WheelEye** subassembly and select **Make Subassembly Flexible**  (*now try to rotate the wheel*)
- Change the appearance of the **BlockForWheel** in the assembly to **Green LED**.
- Complete this layer as shown in Figure 7.40. Again, all parts should be fully mated.
- Add two more layers of yellow blocks as shown in Figure 7.41

Figure 7.40

Figure 7.41

- **Save your work.**

➢ Select **Window, Viewport, Four View** , and your minion should look very similar to the Figure 7.42.

Figure 7.42

Assembly Configurations

- Open your default tall block.
- Select **File, Save As**.
- Use the **Save as copy and open option**.
- Save the new part as **TallBlockHand**.
- Close your tall block.
- Delete all of the configurations for the **TallBlockHand** part, except for the default.
- Delete all of the suppressed features.
- Select the **FrontMidplane** and add the **10 mm** wide feature as shown in Figure 7.43 and the relationships as shown in Figure 7.44.

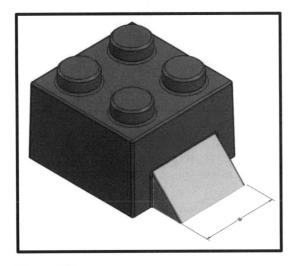

Figure 7.43

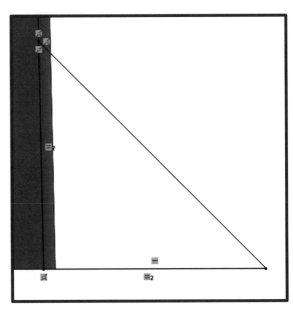

Figure 7.44

- Add a **0.25 mm** fillet as shown.
- **Save your work.** Then close this part.
- Add a second configuration to the assembly, the same way you created a part configuration and name it **TwoEyed**.

Replace Component

- Right click on one of the two black hand blocks in the graphics area and select **Replace Components** 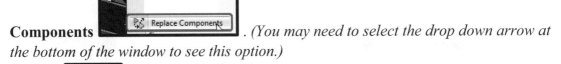. *(You may need to select the drop down arrow at the bottom of the window to see this option.)*
- **Browse** to your **TallBlockHand** part and select **Open**.

➢ Accept this selection and accept the replacement of the mates. Your screen should now look similar to Figure 7.45.

Figure 7.45

To see this being done, watch the video "**Replace Component**."

Suppressing Assembly Components

➢ In the feature tree, select the **WheelEye** subassembly.

➢ **Suppress** this subassembly
➢ Also **Suppress** the two black blocks on each side of it so that your part now looks like Figure 7.46.

Figure 7.46

➢ Insert a new copy of each of the suppressed blocks and two copies of the suppressed assembly and fully mate them. Your **TwoEyed** configuration should look similar to Figure 7.47.

Figure 7.47

Verify that both configurations work by selecting them one at a time. When you suppress a part, you also suppress its mates, and this may have unexpected consequences. Try to move around blocks on this level and the one above. If you can move anything, then add mates as required.

➢ You will need to suppress these three newly inserted parts in the default configuration.

➢ **Save your work.**

Pack and Go

If you wish to move parts to a new computer, you can copy them using Windows Explorer. However, as soon as you wish to move an assembly and/or a part and a drawing, you can no longer do this through Windows Explorer.

All of your parts are linked to their drawing and any assembly they are used in. These links mean that if you change a feature size in a part or in an assembly or in a part drawing, it is also changed in the other two. If you copy a part using Windows Explorer, the links will be broken and must be repaired if you want to see your part.

Pack and go will ensure that all parts, assemblies, fasteners, and drawings are all moved properly.

➢ Select **File, Pack and Go** `Pack and Go...`

Chapter 7: LEGO Minions, Part 2—Assemblies

➢ In the **Pack and Go** window, select the following *(even though you do not have some of the components in this assembly, this still is a good practice to follow)*:
- o **Include drawings** ☑ Include drawings
- o **Include Toolbox components** ☑ Include Toolbox components
- o **Include custom decals, appearances and scenes**:
  ☑ Include custom decals, appearances and scenes

➢ Select **Browse** [Browse...] and navigate to a memory stick to use as a backup for all your work. *(You can also choose to save it as a zipped file.)*
➢ Select **OK.**
➢ **Save** [Save]

Exploded Assembly

➢ Open your **WheelAndHub** assembly.
➢ Select the configuration tab.
➢ Right click on the **Default** configuration and select **New Exploded View**:
[New Exploded View...]
➢ Select the **wheel**. (*For SolidWorks 2015:* please refer to page 156 and then return here.)
➢ Select the blue arrow pointing out of the center of the wheel as shown in Figure 7.48.

This will be the direction of the explosion. If it is pointing in the opposite direction, you can reverse it 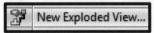 *by selecting the reverse direction option in the Explosion properties window.*

Figure 7.48

➢ In the distance field, enter **20**  *(or drag the blue arrow until you are satisfied with it position).*
➢ Accept this setting and exit the explosion property manager.

➢ Select the **Exploded Line Sketch** tool [Explode Line Sketch]

- Select the inner diameter of the hub as shown in Figure 7.49.

Figure 7.49

If the arrow is pointing away from the tire, then turn on the reverse direction option 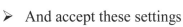 *or click on the arrow when it turns orange* .

- Select any of the internal cylindrical faces on the tire.
- And accept these settings
- Exit the exploded line sketch property window.

- Right click on the **ExpView1** and select **Collapse**
- Right click on the **ExpView1** and select **Explode**.

*(To animate the collapse or the explosion, select the **Animate** option when right clicking.)*

To see the exploded assembly being created, watch the video "**Explode Wheel**."

- Close this assembly.
- Create an explosion for your **WheelEye** assembly that looks like Figure 7.50.

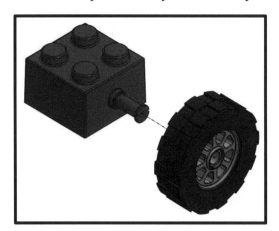

Figure 7.50

Chapter 7: LEGO Minions, Part 2—Assemblies **155**

➢ Create an exploded view of each of your minions as shown in Figures 7.51 and 7.52.

Figure 7.51

Figure 7.52

➢ Explode each layer by **10 mm**. *(Try window selecting your components from the top left.)*

➢ After each step select **Apply** and **Done**

156   Chapter 7: Lego Minions, Part 2—Assemblies

SolidWorks 2014 users can skip this page.

2015 Explosions

In SolidWorks 2015: you have two choices for exploded steps:

1) The **Regular step** , which involves translation and rotation.

2) The new **Radial step** option, which allows the explosion of radially/cylindrically aligned components in one step.

For a demonstration of this step, watch the video "**Radial Explode**."

➢ Select the **Regular step** and return to page 153.

8: LEGO Minions, Part 3—Drawing

➢ Open your default minion assembly.

➢ Select **File, Make Drawing from Assembly**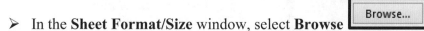

➢ In the **Sheet Format/Size** window, select **Browse**

➢ Navigate to your **$SolidWorksSetup** folder and open it.

➢ Select your **ANSI-D-SIZE-MM** sheet

➢ Select **Open**.
➢ Select **OK**.

➢ **If you set up the custom templates,** then you may see a different screen when you start a drawing. You may see something similar to Figure 8.1.

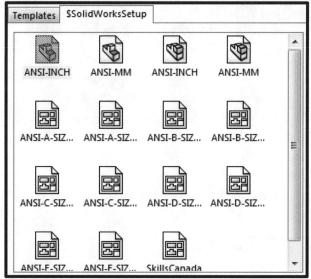

Figure 8.1

From this window, you can select the drawing sheet type and size. The first four templates shown are your part and assembly templates, not drawing sheet templates.

157

➢ The **View Palette** will open up on the right side of the screen as shown in Figure 8.2

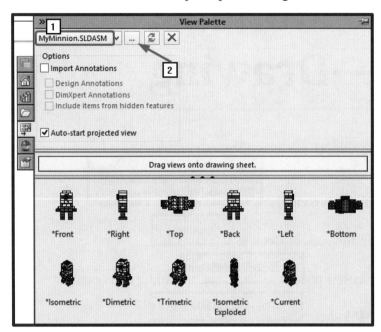

In area 1 of Figure 8.2, the name of the part or the assembly that you are currently creating a drawing for is displayed.

In area 2 is a button that will allow you to select a different part or assembly.

*When adding a view through the view palette, if you check the box for **Import Annotations**, then your dimensions should appear after you drag and drop the view onto the sheet.*

Figure 8.2

➢ Click and hold on the isometric view in the view palette and drag and drop this view into the drawing sheet.
➢ In the **Drawing View Property Manager**, from the **Display Style** section, select the **Shaded with Edges** option  and **Close** the dialog
➢ **Save your work.**

To see this being done, watch the video "**Starting a Drawing**."

View Placement

➢ Move your mouse near your assembly until the drawing view is outlined by an orange line as shown in Figure 8.3.

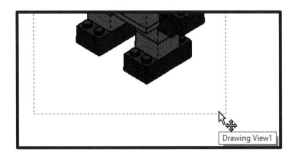

Figure 8.3

➤ Click and hold and move your drawing view so that your view is in approximately the same location as that shown in Figure 8.4.

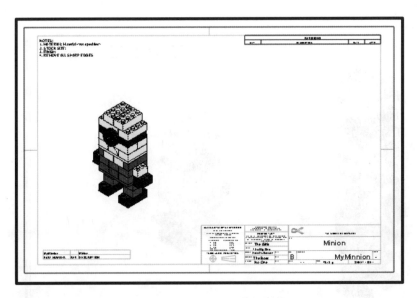

Figure 8.4

Exploded View

➤ Position your mouse near the view again until the orange outline appears around it and click. This will open up the **Drawing View Property Manager** (*typically on the left side of the screen*).
➤ In the **Reference Configuration** section, check the box to **Show in exploded state:**

➤ Accept this selection by closing the dialog
➤ Reposition your view so that it is roughly centered vertically on the drawing sheet.
➤ From the **View Pallet** , drag and drop the isometric view onto the sheet again.
(Notice that the isometric view has a picture of a drawing sheet with a green check mark 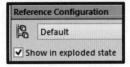 *. This is letting you know that the view has already been placed on the drawing sheet as a base view.)*

➢ Make the display style shaded with edges and close the dialog. Your drawing should look similar to Figure 8.5.

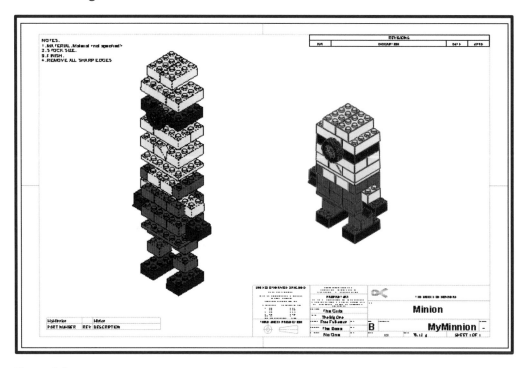

Figure 8.5

➢ Right click on your sheet in the white space. Select **Edit Sheet Format**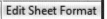
➢ Select the notes in the upper left-hand corner and delete them.

➢ Return from editing your sheet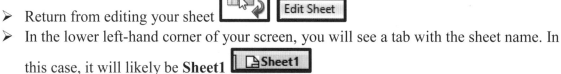
➢ In the lower left-hand corner of your screen, you will see a tab with the sheet name. In this case, it will likely be **Sheet1**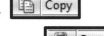
➢ Right click on this tab and select **Rename**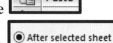
➢ Enter **OneEyeAssembly** as the name and press the **Enter** key on the keyboard when done.

➢ Right click on the sheet name tab and select **Copy**

➢ Right click on the sheet name tab again and select **Paste**

➢ In the **Insert Paste** window, select **After selected sheet**
➢ Select **OK**. *(You have just created a sheet set; look at the lower right-hand corner of the title blocks.)*
➢ Rename this second sheet **TwoEyedAssembly**.

Chapter 8: LEGO Minions, Part 3—Drawing

> Select each view, one at a time. In the **Reference Configuration** menu, select the drop down arrow and choose the **TwoEyed** configuration . *(Remember to make the left side drawing view the exploded configuration.)*

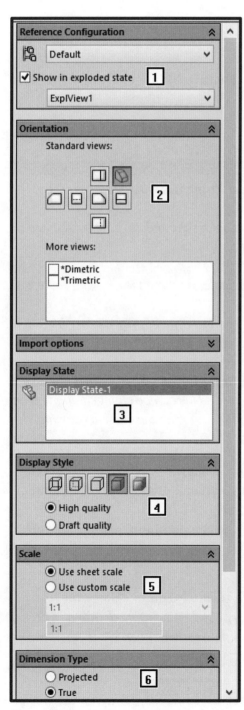

Figure 8.6

Drawing View Property Manager

As shown in figure 8.6, there are four main areas to the ***Drawing View Property Manager:***

Section 1, the **Reference Configuration** *area, allows you to choose which part or assembly configuration to display. In the case of assemblies, it allows you to choose the exploded or collapsed state.*

Section 2, **Orientation,** *allows you to choose which of the standard views you want.*

Section 3, **Display State,** *allows you to choose the display state.*

Section 4, **Display Style,** *lets you choose whether the view should be high or draft quality. The display styles are the same as that of the part and assembly. These are (left to right): wireframe, hidden lines visible, hidden lines removed, shaded with edges, and shaded.*

Section 5, **Scale,** *allows you to set the view to the default sheet scale or to use a custom scale for the view.*

Section 6, **Dimension Type,** *allows you to choose between true or projected dimension styles*

162 Chapter 8: LEGO Minions, Part 3—Drawing

A Warning About Custom Scales: I would never encourage using a custom scale, unless it is for an isometric view. It is not uncommon for a machinist working from our part drawings to use a pocket ruler or scale to measure a part drawing to find a reference dimension. If you display the view in a scale that is not the sheet scale, many unforeseen problems can arise when your part is manufactured. The custom scale is a great tool for deciding what scale to use to make the best use of the drawing sheet. But once you have decided on what scale to use, change the scale setting back to the sheet scale and then alter the sheet scale.

Feature Manager Design Tree

➢ Right click on **OneEyedAssembly** sheet name in the **Feature Manager Design Tree** and select **Properties**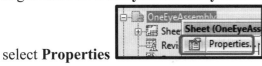

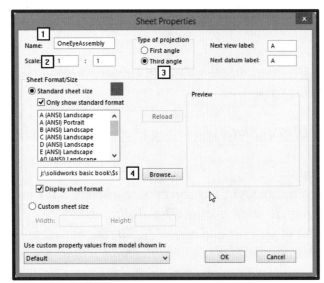

As shown in Figure 8.7, there are four main areas to the **Sheet Properties** window.

Section 1 displays the sheet **Name**, which can be edited here.

Section 2, **Scale**, is where the drawing scale is set.

Section 3 sets the **Type of projection**, First angle or Third angle.

Section 4, **Sheet Format/Size**, is where the sheet standard is set. This can be changed at any point in time without altering the current drawing views.

Figure 8.7

➢ Expand the sheet by selecting the plus (+) sign beside the name.
➢ Expand a drawing view. You will notice that the entire part and/or assembly feature tree is present and features can be shown or hidden from here.
➢ Right clicking on any of the drawing views will provide the option to **Open** the part or assembly

Just like in the part feature tree, you can select a sheet here and drag it to change the order in the sheet set. Any drawing view can also be selected and dragged and dropped onto a different sheet. If you right click on a sheet in this area, you can also add another sheet. Or if you perform a slow double click on a sheet, you can rename it.

➢ Right click on one of your sheets and select **Add Sheet**. *(You may have to browse to your sheet, but only the first time you add a sheet. From that point forward, SolidWorks will automatically add that sheet style for each new sheet inserted into this sheet set.)*

Once you place your part or assembly into a drawing, the title block information will populate. This information is coming from the properties you set up in your part. To change anything in your title block, edit the properties of the part or the assembly. If you do not recall what property is linked from your part or assembly to the cell in the title block, right click in the white space of the drawing sheet and select **Edit Sheet Format**.

If the links are missing, you can go back and repeat the steps from pages 29 to 32. Alternatively, you can manually enter or link the information while editing the sheet. While the sheet is being edited, all drawing views will not be visible.

Restoring Broken Title Block Link

> Right click in the white space of the drawing sheet and select **Edit Sheet Format**

> From the **Annotation** tab, select **Note**. *(Do not worry about the leader or the placement, these will be fixed later.)*

> Place the box approximately in the required field. *(For instance, if your name is missing from the* **Drawn By** *field, then place the box in this area of the title block.)*

> On the left side of the screen, you will find the **Note Property Manager**.

> In the **Text Format** section, select **Link to Property**

> In the **Link to Property** window, select the **Model in view specified in sheet properties** radio button

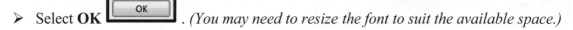

> From the drop down menu, select the desired property

> Select **OK** . *(You may need to resize the font to suit the available space.)*

> If the note is still active, press the **ESC** key on your keyboard.

> Alternatively, you can select the **File Properties** option.

> This will open up the current sheets properties, and you can add the desired property.

> Select **OK**.

> Then select the property from the drop down list, with the **Current Document** option selected

> Return from editing your sheet

To see this being done, watch the video "**Restoring Title Block Links**."

164 Chapter 8: LEGO Minions, Part 3—Drawing

Standard Three View

- Rename your third sheet **Tall Block.**
- Select the **View Layout** tab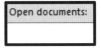
- Select **Standard 3 View**

- If your tall block is currently open, it will appear in the **Open documents** section. Select it.
  - At this point, you can also select the **Reference Configuration** to open:

- If the tall block part is not open, select **Browse** and navigate to the folder with your part by following these steps:
  - In the **Open** window, near the lower right-hand corner, select the drop down arrow for file type 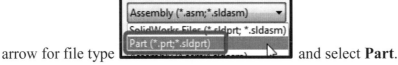 and select **Part**.
  - Select your **TallBlock**.
  - Select **Open**

Your drawing should look similar to Figure 8.8.

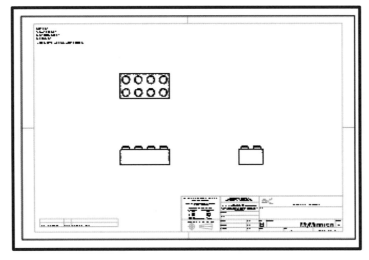

Position your mouse near one of the parts, until the orange rectangle appears around the view.

Click and drag the view around the sheet. Do this for all three views, and notice how each view is aligned with the other.

Figure 8.8

- Select the front view.
- In the **Property Manager**, **Reference Configuration,** choose the **Default** configuration.
- Do this again for the top and right side view.

Selecting a Scale

➢ Select the **Front** view.

➢ In the **Property Manager, Scale** section, choose **Use custom scale**:

➢ Select the **4:1** scale. Your drawing should look similar to Figure 8.9

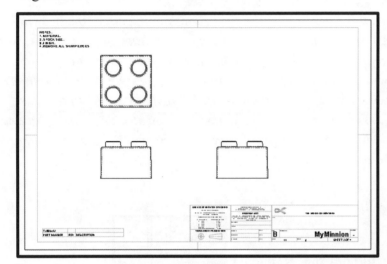

Only use the custom scale option to help you choose what sheet scale you would like to use.

Figure 8.9

➢ Switch the scale option to **Use sheet scale**
➢ Right click in white space on the drawing sheet or on the **Tall Block** tab across the bottom or on the **Tall Block** sheet name in the property manager and select **Properties**

➢ Set the **Scale** to **4:1** and select **OK**.

Deleting a View

➢ For a part such as this, the front and right side view are the same. Since there is no need to duplicate a view, position your mouse near the right side view. Once it is selected

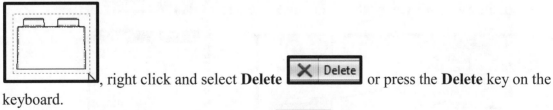, right click and select **Delete** or press the **Delete** key on the keyboard.

➢ In the **Confirm Delete** window, select **Yes**

Showing Hidden Lines

➢ Select your front view

➢ Select the **Hidden Lines Visible Display Style** , your drawing should now look similar to Figure 8.10.

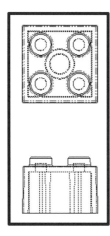

Both of your views will have hidden lines. This is because you selected the base view (the first view you placed on the drawing sheet). If you only want to display hidden lines in the top view, then select it and then turn on the hidden lines display option. When showing hidden lines, you should always ask yourself, "Do they aid in showing the details of the part with clarity or not?" In this case, I would suggest that they make the drawing slower to interpret and more difficult to interpret. A section view would be a better choice to add clarity.

Figure 8.10

➢ Set the display back to **Hidden Lines Removed**

Cross Section

➢ Select the **View Layout** tab
➢ Select the **Front** view.

➢ Select **Section View**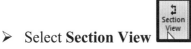
➢ Use the **Vertical** cutting line and place it on the midpoint of the line as shown in Figure 8.11 and accept these selections.

➢ In the **Section View** window (Figure 8.12) select **OK**

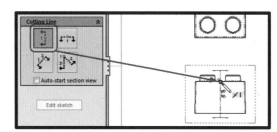

Figure 8.11

Figure 8.12

Chapter 8: LEGO Minions, Part 3—Drawing **167**

➢ Position your view to the right of the front view and click to place it.

➢ In the **Section View Property Manager** set the section **Label** to **A** and accept these selections . Your drawing should look similar to Figure 8.13.

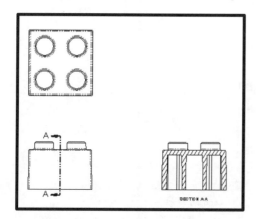

Figure 8.13

For a more detailed explanation of how to create section views, watch the video "Creating Section Views."

Inserting Solid Creation Dimensions

➢ Select **Annotation**

➢ Select **Model Items**.

➢ In the **Model Items Property Manager**, **Source/Destination** section, select **Entire model** as the the source for the dimensions. *(The model items are any dimensions used in sketches and in features to create your part.)*

Chapter 8: LEGO Minions, Part 3—Drawing

➢ Turn on the **Import items into all views** option. Do not change any settings. Accept these selections. Your drawing should now look similar to Figure 8.14.

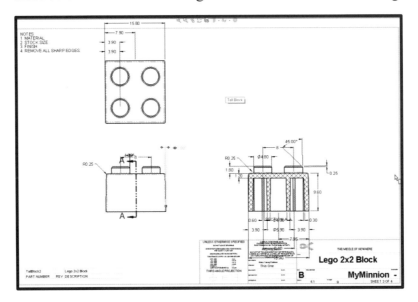

Figure 8.14

Your dimensions very likely will need to be moved so that they are neatly placed between the views on the drawing sheet. The extension lines may also require some adjustment, as they should not overlap an object line. You may also wish to move a dimension from one view to another.

➢ Click and hold on a dimension that you wish to move and then drag it to the desired location, releasing the mouse when you are done.

To see the dimensions being neatened up, moved and/or copied, watch the video "<u>Fixing Dimensions</u>."

Moving/Copying a Dimension

To move a dimension from one view to another, hold down the **Shift** key while clicking and holding on the dimension to move. Drag the dimension to the new view and release the mouse button first.

To copy a dimension from one view to another, hold down the **CTRL** key instead of the **Shift** key.

While you are moving the dimension, you should notice a red circle [icon] with a line through it on the lower right-hand side of your cursor. This means that you cannot move or copy the dimension to the current location. Once this symbol changes to an image of a black object line with a blue dimension [icon], then you can release the mouse to place the dimension.

Neatening Dimensions

After you move a dimension's original placement on a drawing view, the extension lines may be on top of the object lines, or they may have too large of a gap. To fix either problem, select the dimension. The extension lines and dimension lines will now be blue. Click on and hold the blue box at the end of the extension line, drag it to the desired location, and then release the mouse button to neaten up the dimension.

Leaders can be moved from one location on a drawing view to another by using the same method. Select the dimension, click and hold on the blue end of the leader, drag it to a similar feature of the same size, and release the mouse to place it at the desired location. *(The leader can only be moved to a feature of the same exact size)* Once the desired edge turns orange, the mouse button can be released to place the dimension. Any drawing item, a section caption, a section line, a note, and so on, may be moved in this manner.

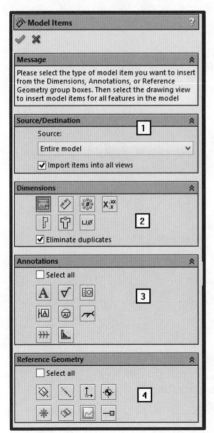

Model Items Property Manager

As shown in Figure 8.15, there are four main areas in the ***Model Items Property Manager****:*

Section 1, ***Source/Destination,*** *allows you to choose to import the dimensions into all views or just the selected view, as well as to import dimensions from the entire model or just from the selected feature in the view.*

Section 2, ***Dimensions,*** *allows you to choose what type of dimensions will be imported as well as the elimination of duplicate dimensions. To add any of these dimension types to your drawing simply select the desired icon (see the list of icons below).*

Section 3, ***Annotations,*** *allows you to select model annotations to be added to the drawing (see the list of icons below).*

Section 4, ***Reference Geometry,*** *allows you to add reference geometry to your drawing (again, see the list of icons below).*

Figure 8.15

Dimensions Icons

 Dimensions marked for the drawing. This is the default setting, and all dimensions used to create sketches and feature are marked for the drawing unless you unmark them.

 Dimensions not marked for the drawing.

 Instance counts for circular sketch or feature patterns.

 Tolerance dimensions.

 Hole wizard profiles.

 Hole wizard locations.

 Hole callouts.

Annotations Icons

 Notes.

 Surface finish.

 GD&T (Geometric Dimensioning and Tolerancing).

 Datums.

 Datum targets.

 Weld symbols.

 Caterpillar weld information.

 End treatment of weld information.

Reference Geometry Icons

 Planes.

Axis.

Origins.

Center of mass.

 Points.

 Surfaces.

Curves.

 Routing points.*

* Routing points are often used when adding features such as hydraulic or pneumatic tubing.

If you do not have all of the dimensions shown, then add them using the Selected feature Selected feature *option from the insert model items command. Or use the smart dimension tool to add them as you dimensioned features in a sketch.*

> Neaten up your drawing by rearranging, moving, and/or adding dimensions by employing the **Model Items Property Manager**, so that it looks similar to Figure 8.16.

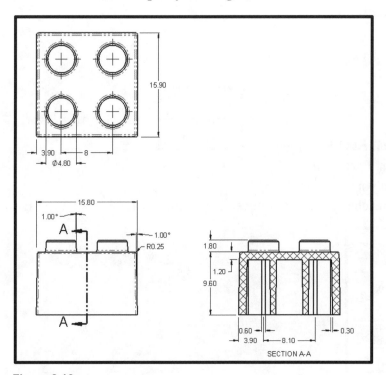

Figure 8.16

Adding Center Mark/Line

> To add a center mark, from the **Annotation** Annotation tab select **Center Mark**: Center Mark
> Click on any circle in the drawing to add a center mark as shown in Figure 8.16 above and exit the command when done.

> To add a center line, from the **Annotation** tab select **Centerline** Centerline
> Select two entities that you wish to place the centerline between.

Switching to a Leader Diameter Dimension

> Right click on the **Ø4.80** dimension and select **Display Options** Display Options
> Then select **Display As Diameter** Display As Diameter
> Move this dimension so that your drawing now looks similar to Figure 8.17.

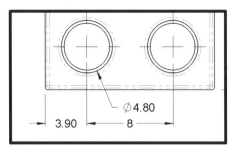

Figure 8.17

Adding Text to a Dimension

- Select (*single click on*) the **Ø4.80** dimension
- Click in the **Dimension Text** field.
- Press the **Home** key on your keyboard.
- Enter **4 x** and accept this change. Your dimension should now look similar to Figure 8.18.

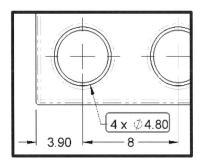

Figure 8.18

To see this being done, watch the video "**Editing Dimension Text**."

Chamfer Dimension

- Select the **Annotations** tab
- Select the drop down arrow below **Smart Dimension**
- Select **Chamfer Dimension**

Chapter 8: LEGO Minions, Part 3—Drawing

➢ Select the two entities which the chamfer is comprised of, as shown in Figure 8.19, and click to place the dimension text in the desired location.

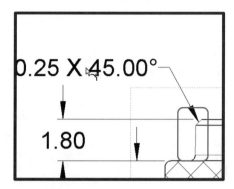

Figure 8.19

To see this being done, watch the video "**Adding a Chamfer Dimension**."

Setting Dimension Precision

➢ Select the chamfer dimension.
➢ In the **Dimension Property Manager** *(recall: the right side of your screen is the default location for this to show up)* from the **2nd Tolerance/Precision** section, select the drop down arrow for the **Unit Precision** field *(the lower box)*, set the precision to **None,** and accept this selection. Your dimension should now look similar to Figure 8.20.

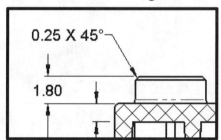

Figure 8.20

To see this being done, watch the video "**Setting Dimension Precision**."

Adding a Note

- Delete the two **1.00 degree** dimensions and the **R0.25** dimension.
- Right click on the drawing sheet in white space and select **Edit Sheet Format**:

  [Edit Sheet Format]

- Double click on the **Notes** section in the upper left hand corner of the sheet.

All of your drawing views will temporarily disappear while you are editing the sheet.

- Click inside of the textbox
- Delete the **STOCK SIZE** and the **FINISH** labels
- Add the following two notes:
  - **3. ALL DRAFT ANGLES ARE 1 DEGREE**
  - **4. ALL FILLETS AND ROUNDS ARE 0.25 UOS**
- Click outside of the text box to stop editing.

- Stop editing the sheet format . Your notes section should look similar to Figure 8.21.

```
NOTES:
1. MATERIAL:
2. REMOVE ALL SHARP EDGES
3. ALL DRAFT ANGLES ARE 1 DEGREE
4. ALL FILLETS AND ROUNDS ARE 0.25 UOS
```

Figure 8.21

To see this being done, watch the video "**Editing Your Drawing Notes**."

Dimension Property Manager

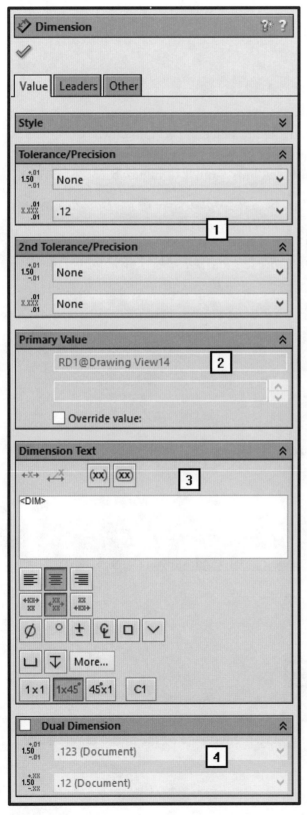

Figure 8.22

The dimension properties can be edited in the drawing or in the mode. Regardless of where you edit them, both will be updated, because they are linked.

As shown in Figure 8.22, there are four main areas in the dimension property manager:

Section 1, **Tolerance/Precision**, allows you to set the **Tolerance Type** (the upper drop down section) and the **Unit Precision** (the lower) for each part of the dimension. In the case of a chamfer dimension, there will be two **Tolerance/Precision** fields, one for the chamfer size and one for the chamfer angle.

The unit precision controls the number of decimal places.

The tolerance types are: none, basic, bilateral, limit, symmetric, MIN, MAX, fit, fit with tolerance and fit (tolerance only).

Section 2, **Primary Value**, allows you to rename the dimension, shows you what sketch the dimension is part of, and will allow you to change the feature size or to override the actual value.

Section 3, the **Dimension Text** field, allows you to add to the dimension by typing new values, such as "4 x," or by adding common dimension symbols.

Section 4, **Dual Dimension**, allows you to display both inch and metric sizes on your drawings—or any other combination of dimensions, such as inch decimal dimensions and inch fractional dimensions.

176 Chapter 8: LEGO Minions, Part 3—Drawing

Projected View

- Select the front view.
- Select the **View Layout** tab
- Select **Projected View**
- Move the cursor to below the front view and click to place the bottom view.
- Move the cursor to the right of the front view and click to place the isometric view.
- Accept these selections. Your screen should now look similar to Figure 8.23.

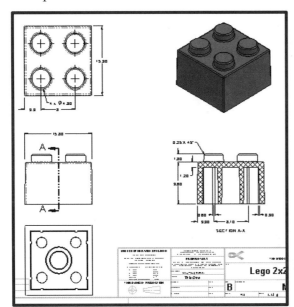

Set the display style of the isometric view to "shaded with edges."
  Notice that if you move your mouse to the left, the left side view will appear, or to the right the right side view will appear.

Figure 8.23

To see this being done, watch the video "**Creating a Projected View**."

Breaking and Restoring the View Alignment

- Select the bottom view.
- Right click and select **Alignment**
- Select **Break Alignment**
- Click and hold on the bottom view, then drag it to the left side of the part and release the mouse.
- Select the view again.

Chapter 8: LEGO Minions, Part 3—Drawing **177**

➢ Right click and select **Alignment, Align Horizontal by Origin**:

➢ Select the front view to align the bottom views origin with the front views origin. Your drawing should now look similar to Figure 8.24.

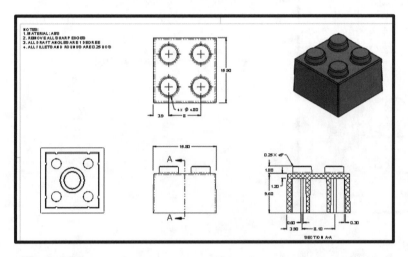

Figure 8.24

To see this being done, watch the video "**Breaking the View Alignment**."

Inserting Dimensions into One View

➢ Select the bottom view.
➢ Select **Model Items** from the **Annotations** tab.
➢ Set the **Source** to the **Entire Model**
➢ Make sure that the **Import items into all views** 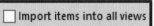 is not checked.
➢ Accept these selections

➤ Add center marks, center lines and neaten up the dimensions so that your view is similar to Figure 8.25.

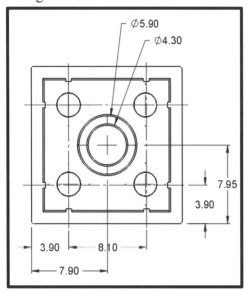

When we create parts, we use the quickest possible solution. However, when we create drawings and insert the dimensions, there may not be enough information for the manufacture of the part. You may need to add more dimensions in order to provide all of the required information for manufacturing.

Figure 8.25

Smart Dimensioning a Drawing

➤ Use the **Smart Dimension** option to dimension your drawing, the same as you would use it to dimension a sketch. Alter all dimensions to look like Figure 8.26.

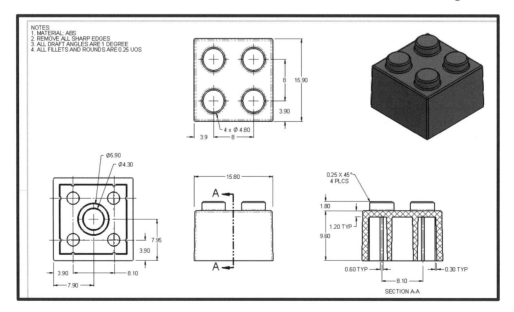

Figure 8.26

➢ Add a sheet for each of the following parts and create the views as shown in Figures 8.27 through 8.30.
➢ Rename each sheet accordingly.

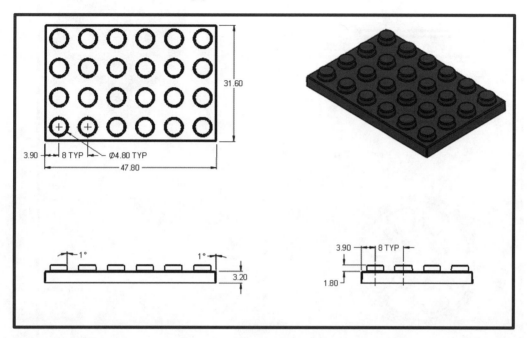

Figure 8.27

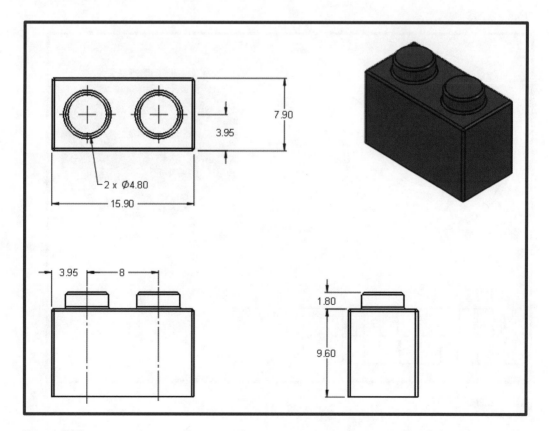

Figure 8.28

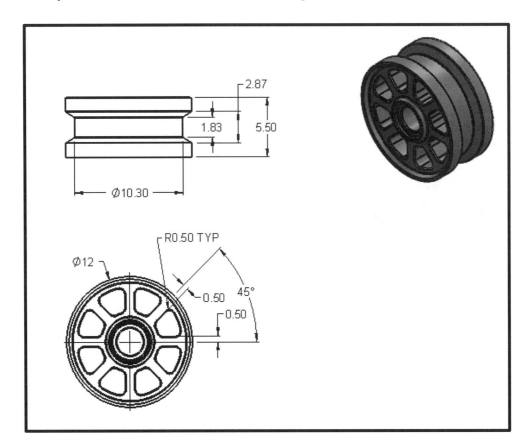

Figure 8.29

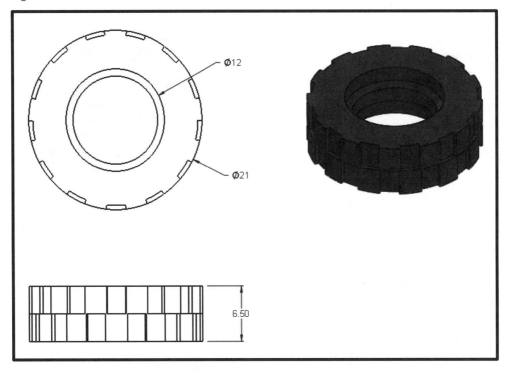

Figure 8.30

> **Save your work.**

If you look in the lower right hand corner of your drawing sheets, the title block should indicate that your sheet set now has seven sheets

Broken-Out Section

> Select your **Short Block** drawing sheet.

> From the **View Layout** tab select **Broken-out Section** . *(The spline sketching tool will be active. Move your mouse around the area where you would like the spline, clicking as you go to place points on the spline. Add the last point over the first to end the spline and then exit the spline tool .)*

> Sketch a spline, in your front view, similar to the one shown in Figure 8.31.

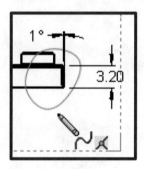

Figure 8.31

> In the **Broken-out Section Property Manager**, set the depth to **3.9 mm** and accept this selection. Your drawing should now look similar to Figure 8.32.

> Add the dimension as shown in Figure 8.32.

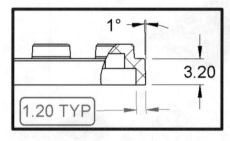

Figure 8.32

To see this being done, watch the video "<u>Broken-out Section</u>."

182 Chapter 8: LEGO Minions, Part 3—Drawing

Adding a Note

> From the **Annotation** tab, select **Note**
> Click below the approximate center of your isometric view to place the text box for the note.

> In the **Note Property Manager,** select **Link to Property**
> In the **Link to Property** window, select **Model in view to which the annotation is attached**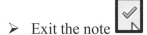
> From the drop down menu, select **SW-Configuration Name (Configuration Name)**

> Select **OK**

> Highlight the **4x6** text and make it **Bold**

> **Close** the text formatting window

> Exit the note

To see this being done, watch the video "**Adding a Linked Note**."

> Add the views, dimensions, and note to your part drawing so that your sheet now looks similar to Figure 8.33.

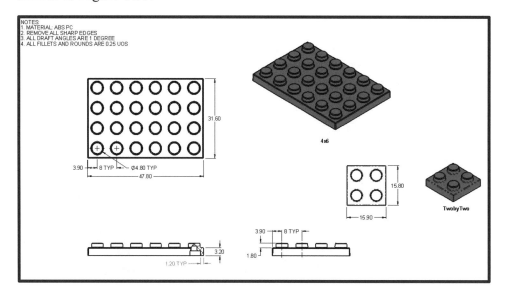

Figure 8.33

Detail View

- From the **View Layout** tab, select **Detail View** . *(Your cursor will now show the circle sketching icon 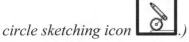.)*
- Sketch a circle as shown in Figure 8.34. (*Click on your drawing to place the circle center and then move your mouse and click to set the circle radius. Be certain that all the features that you wish to show in detail are contained in the circle.*)
- Position your mouse on the sheet where you would like to place the view and click.
- Make this view **B** and accept the default settings in the **Detail View Property Manager**.
- Add the chamfer dimension and set the detail views scale to **6:1** as shown in Figure 8.34.

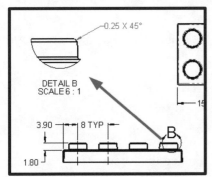

Figure 8.34

The detail view label, in this case the letter B, will be added to the circle as soon as you finish creating the circle.

To see this being done, watch the video "**Creating a Detail View**."

- Edit your **OneByBlock** sheet so that it looks similar to Figure 8.35

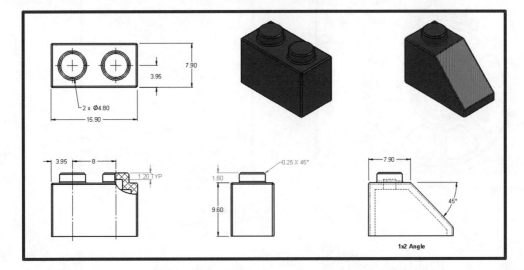

Figure 8.35

Aligned Section View

> Select your **Hub** sheet.

> From the **View Layout** tab, select **Section View**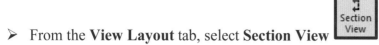

> In the **Section View Assist Property Manager**, **Cutting Line** section, choose the **Aligned** option

> Click on the center of the hub in the front view to lock the rotation point.
> Select the midpoint of the arc, as shown in Figure 8.36, as the location to lock the first arm of the section view
> Select the left endpoint of the horizontal center to lock the second arm of the section

> Accept these selections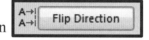
> Move your mouse below the front view and click to place this view.
> Make the section view **C-C** and set the direction so that your screen looks similar to Figure 8.37. You may need to use the **Flip Direction** button

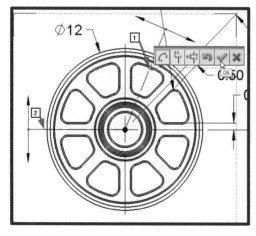

Figure 8.36

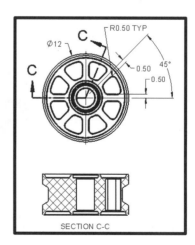

Figure 8.37

To see this being done, watch the video "**Creating an Aligned Section**."

Dimensioning to a Virtual Sharp

- Activate **Smart Dimension**.
- In your section view, right click on the orange-colored line, as shown in Figure 8.38, and select **Find Intersection**.
- Now select the top horizontal line as shown in Figure 8.39.

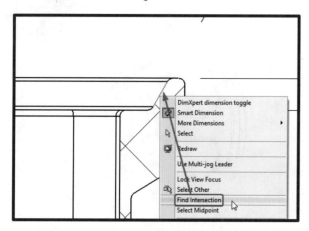

Figure 8.38

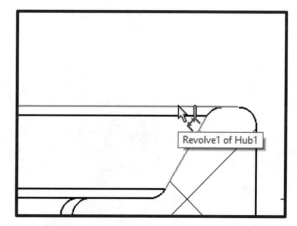

Figure 8.39

- The **Virtual Sharp** will now appear as shown in Figure 8.40.
- Repeat the procedure for the opposite side of the part, selecting the two lines as shown in Figure 8.41.

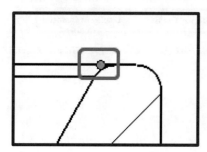

Figure 8.40

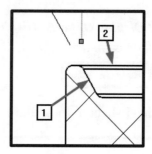

Figure 8.41

- Move the mouse up and place the dimension by clicking at the desired location.

- Add the **Diameter** symbol to the dimension by selecting it from the **Dimension Text** section of the **Dimension Property Manager** and close the dimension option.

To see this being done, watch the video "**Dimensioning to a Virtual Sharp**."

➤ Edit your **Hub** sheet so that it looks similar to Figure 8.42.

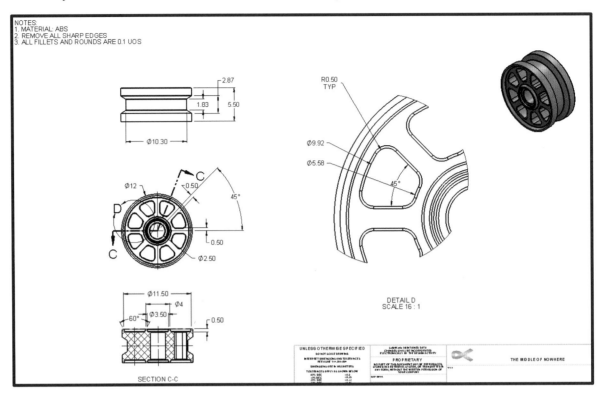

Figure 8.42

To change the display of a radius dimension to a diameter dimension, right click on the dimension and select **Display Options, Display as Diameter**. *To display a diameter dimension as a radius, right click the dimension and select* **Display Options, Display as Radius**.

➤ Edit your **Wheel** sheet to look like Figure 8.43.

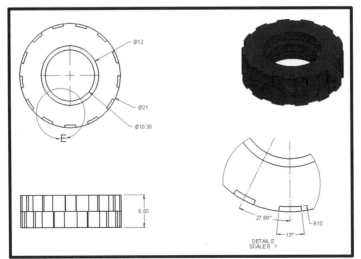

Figure 8.43

At this point, you may have realized that there are several drawings missing from the sheet set. You will now insert them into the correct location.

Reordering the Drawing Sheets

- Add a sheet to the set as you did before.
- In the **Feature Manager Design Tree**, or in the tab section across the bottom, click and hold on the new sheet. Now drag and drop it so that it is the third sheet in the set.
- Rename this sheet **Wheel and Hub**.

To see this being done, watch the video "**Reordering Drawing Sheets**."

Creating Drawing Views Using Model View

- From the **View Layout** tab, select **Model View**

- In the **Model View Property Manager,** select **Browse**
- Browse to the location of your assemblies.
- In the **Quick Filter** section of the **Open** window, select **Assembly**:

- Double click on your **WheelAndHub** assembly.

- In the **Model View Property Manager,** select the **Isometric** view from the **Orientation** section.

- From the **Display Style** section, choose **Shaded With Edges**
- Click to place the view on the sheet.
- Add another isometric view to the sheet. This time, turn on the option to **Show in exploded state**. Your drawing sheet should look similar to Figure 8.44.

Figure 8.44

BOM (Bill of Materials)

➢ From the **Annotations** tab, select the down arrow below **Tables**

➢ Select **Bill of Materials**

➢ When prompted to **Select a drawing view to specify the model for creating a Bill of Materials** 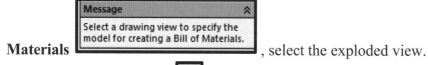, select the exploded view.

➢ Accept the default settings and the BOM should be added to your drawing sheet as shown in Figure 8.45.

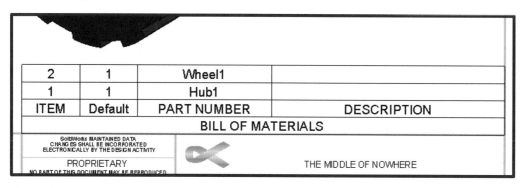

Figure 8.45

To see this being done and for a more detailed explanation about the BOM, watch the video

"Creating a Bill of Materials."

Altering the BOM

➢ Position your mouse just above the description column of the BOM until you see the down arrow and the letter **D** appears

➢ Right click and select **Delete Column**

➢ Position your mouse near the upper left-hand corner of the BOM until the axis and the anchor appear 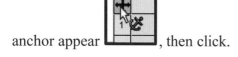, then click.

Chapter 8: LEGO Minions, Part 3—Drawing

➢ In the upper left-hand corner the **Bill of Materials, Property Manager** will appear. At the top of it is the **Table Position** 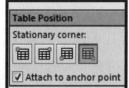. Uncheck **Attach to anchor** point and change the position of your table to the top right-hand side of the drawing sheet.

➢ **Save your work.**

Balloon Items

➢ From the **annotations** tab, select **Balloon**

➢ In the **Balloon Property Manager, Balloon Settings** section, select **Circular** as the style and **1 Character** as the size.

➢ Click on the wheel. Then position the mouse in white space above but near the wheel and click to place the balloon.

➢ Repeat this last step for the rim.

➢ Accept these selections. Your drawing should look similar to Figure 8.46.

Figure 8.46

To see this being done, watch the video "**Adding Balloon Items**."

At any point when working with your drawing, if you notice a bright pink border around a drawing view or around the sheet, you have accidentally locked the focus. If this happens, right click and select **Unlock Sheet Focus**.

Magnetic Line

➢ From the **Annotation** tab, select **Magnetic Line**
➢ Click to the left of the left-most balloon item.
➢ Move your mouse close to the balloon, and it will jump to the magnetic line. Add both balloons to the magnetic line this way.

190 Chapter 8: LEGO Minions, Part 3—Drawing

- Click to end the magnetic line.
- Exit the command
- Select one of the balloon items, and the magnetic line will reappear.
- Click and hold on one of the balloons and drag it off the magnetic line. It will no longer be attached to the line.
- Click and hold on the same balloon item and drag it back to the magnetic line to reattach it.
- Select one of the balloon items again, then grab one of the arrows at the end of the line and drag it around to reposition the balloons.

*Magnetic lines are automatically added when you use the **Auto Balloon** option. They allow you to easily and quickly neaten up a drawing with many balloons.*

To see this being done, watch the video "**Magnetic Line**."

Add another sheet before this one. *(If you ever add a sheet and receive an error message stating the sheet format could not be located, simply close the error message window and then use the browse option to navigate to your sheet.)*

- Name the sheet **WheelEye.**
- Make the sheet look similar to Figure 8.47.

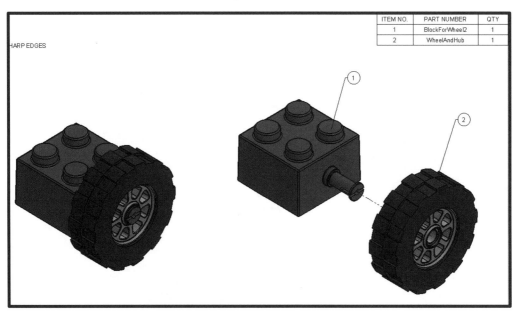

Figure 8.47

Select one of your balloon items, grab the endpoint that is attached to the part, and drag it to a face and release it. It should change to a solid dot . *Now drag it to a part edge, and it should change to an arrow* .

Showing Part Configurations

> Add another sheet, name it **Tall Block Variations,** and set the sheet scale to **1:1**.
> zAdd a top view of your default tall block to the sheet and set the **Display Style** to

Hidden Lines Visible
> Add a second top view of the block. This time, select your 2x3 configuration:
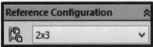
> Add a top view for each of your configurations.
> Add a note below each configuration, linking the note to your configuration name so that your sheet looks similar to Figure 8.48.

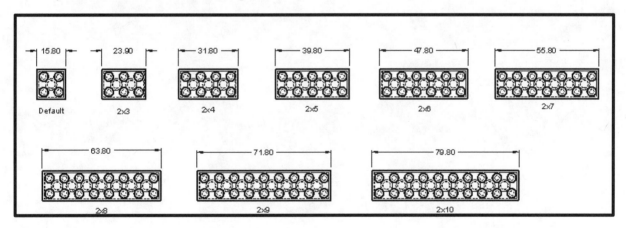

Figure 8.48

Save as a PDF

- Select **File, Save As**
- In the **Save As** window, in the **Save as type** option, choose **Adobe Portable Document Format**
- Select **Save**.
- In the **PDF Export** window, leave all sheets selected and choose **OK**.

9: LEGO Minions, Part 4—Revisions

Tracking design changes is a critical part of a designer's job. Releasing the wrong drawing revision can cost a company an enormous amount of money and valuable customers. Therefore, it is essential that you learn to do this simple task well.

➤ Open your **TallBlockHand** part . If you are still in your drawing sheet, select the isometric view of the part and then click **Open Part**:

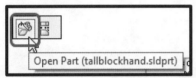

➤ Double click on the angled extrusion to show the dimensions as shown in Figure 9.1.

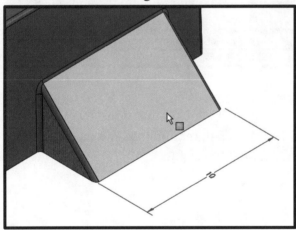

Figure 9.1

➤ Double click on the **10 mm** dimension and change it to **12 mm**
➤ Rebuild the part and accept the dimension change.
➤ **Save your part**.

➤ Rename this feature **Hand Extrusion**

193

194 Chapter 9: LEGO Minions, Part 4—Revisions

➢ Right click on your part name in the feature tree and select **Comment, Add Comment**:

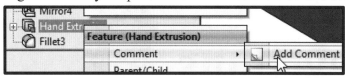

➢ Enter the comment as shown in Figure 9.2 and add the time and date stamp to the comment.

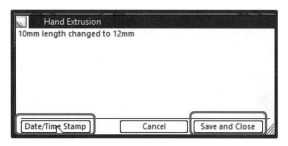

Figure 9.2

➢ **Save and Close** the comment.

➢ Select **File** [File], **Properties** [Properties...]

➢ Add the **Custom** property [Summary | Custom] as shown in Figure 9.3 and select **OK**.

Property Name	Type	Value / Text Expression	Evaluated Value
Revision	Text	Hand Extrusion 10 mm change to 12mm	Hand Extrusion 10 mm chang

Figure 9.3

➢ Close your part and open the drawing. The **Tall Block Variations** sheet should show the design change as seen in Figure 9.4. If it does not, rebuild the drawing. (*If you open your assembly, you will notice that the revised part is there.*)

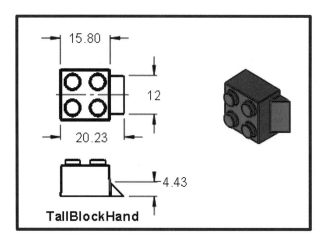

Figure 9.4

If your template does not have a Revision Table, then complete the following steps 1-3. If your template does have a revision table, then skip these steps.

1) Select **Tables**, from the **Annotation** ribbon bar, then select **Revision Table**:

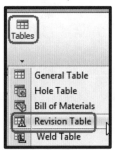

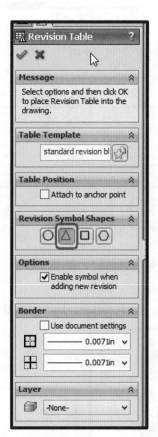

2) Set up the **Revision Table** options as shown

3) Accept these selections . A revision table will appear in the upper right hand corner of your drawing sheet:

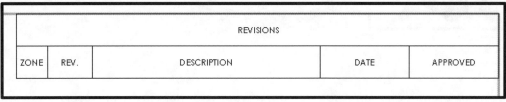

196   Chapter 9: LEGO Minions, Part 4—Revisions

➢ Go to your first sheet, **OneEyeAssembly.**
➢ Right click on the **Zone** column and select **Delete, Column:**

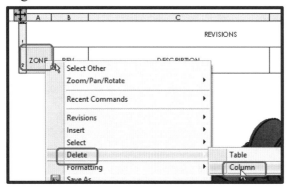

For a large sheet, you will want to identify the zone where the revision is located within the sheet. This will make identification of the change quicker for the reader of the print.

➢ Right click anywhere in the revision table and select **Revisions, Add Revision**

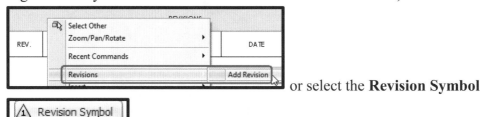 or select the **Revision Symbol** from the Annotation ribbon bar.

➢ Click on the altered part, move your mouse off the part, and click to place the revision symbol near the altered block as shown in Figure 9.5. Press the **ESC** key on your keyboard to exit the **Revision Symbol Property Manager**.

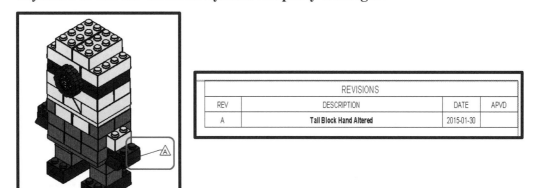

Figure 9.5

Add the revision symbol and table note to both of your assembly drawing sheets.

- Double click in the **Description** cell of the revision table beside the letter **A** and below the word **Description**.
- Enter **Tall Block Hand Altered**.
- Click in white space to end the revision note.
- Select the **Tall Block Variations** sheet.
- Select the **Annotation** ribbon bar and select **Revision Cloud**. *(It is important to clearly identify all revisions and draw people's attention to the changes.)*

- Draw a revision cloud around the new dimension. *(Click above and to the left of the dimension to start the revision cloud and then click below and to the right of the dimension to end it.)* Use any style that you like.
- Exit the **Revision Cloud Property Manager**.
- Add the revision symbol to the revision cloud as shown in Figure 9.6.

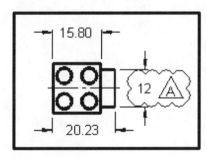

Figure 9.6

- Double click in the **REV** cell and add the letter **A**.
- If there is a note in the **Description** cell, double click in it and delete the note.
- Add the note: **12 mm was 10 mm**.
- Add the date.
- **Save your work**.

10: Bar Puller

A bar puller can be used on a CNC lathe to pull the bar stock out of the chuck, thus allowing multiple parts to be made from one length of stock. Figure 10.1 shows the actual bar puller that you will be drawing.

*Save your parts as **YourName_PartName**.*

Figure 10.1

Shank

➢ Create the **Shank** as shown in Figure 10.2.

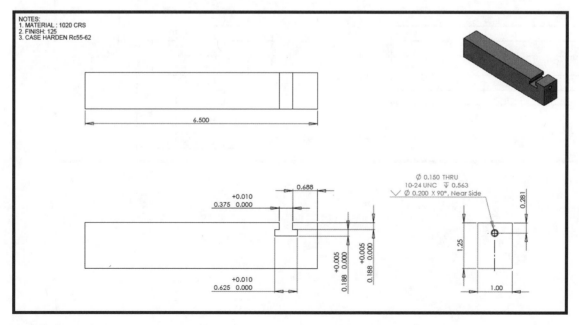

Figure 10.2

Chapter 10: Bar Puller

To see this part being created, watch the video "**BP Shank Sketch**."

To see the hole wizard hole being created, watch the video "**BP Hole Wizard**."

Create Remaining Parts

Create the parts shown in Figures 10.3, 10.4, 10.5, 10.6, and 10.7, using symmetry, where possible, and the appropriate dimension style. This should only take you a few minutes. Unless otherwise stated, make the material **1020 CRS**.

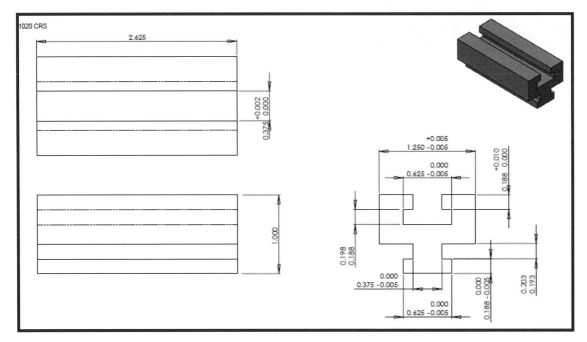

Figure 10.3 (T SLOT)

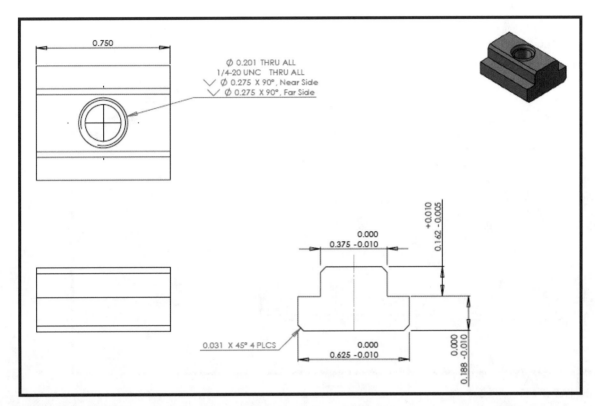

Figure 10.4 (T NUT)

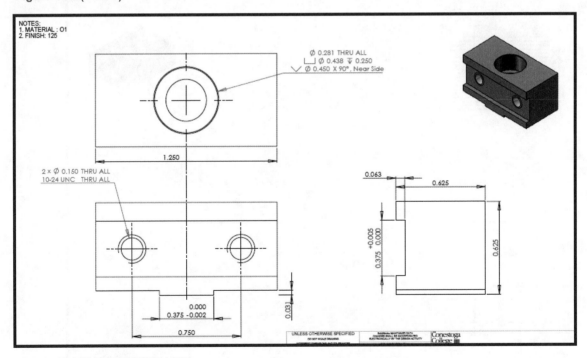

Figure 10.5 (ARM ADJUST NUT)

Chapter 10: Bar Puller

When you create the part shown in Figure 10.5 and 10.6, **do not** include the two tapped holes and the two countersunk holes.

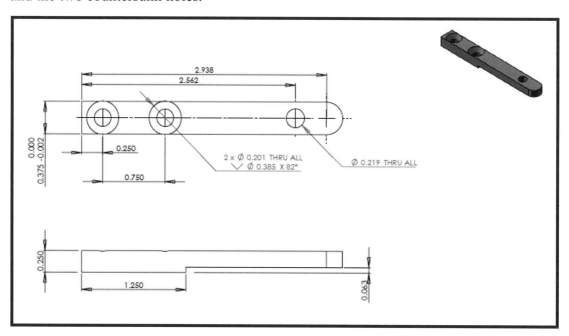

Figure 10.6 (ARM)

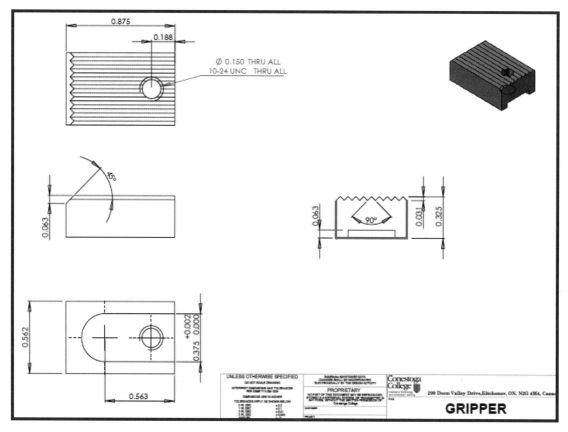

Figure 10.7 (GRIPPER, Material O1)

Toolbox

Before proceeding, check to see if your Toolbox add-in is enabled. *Earlier, you configured the toolbox; it should still be enabled.*

> From the Tools drop down menu, select Add-Ins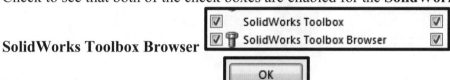
> Check to see that both of the check boxes are enabled for the **SolidWorks Toolbox** and the **SolidWorks Toolbox Browser**
> Select **OK** once they are turned on

Adding a Material

> Right click on **Material** in the feature tree and select **Edit Material:**

> Right click on **Custom Material** and select **New Category**
> Name your new category **Tool Steels.** *(SolidWorks does have some tool steels as part of its library, but not all.)*

> Right click on your new category folder and select **New Material**

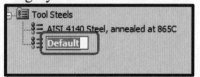

> Name your new material **01.**
> Set up all of the materials properties and **Save** this material.

To see this being done, watch the video "SW New Material."

01 Tool Steel Properties

Property	Value	Units
Elastic Modulus	2.137371e+011	N/m^2
Poisson's Ratio	0.3	N/A
Shear Modulus	8.2e+010	N/m^2
Mass Density	7830	kg/m^3
Tensile Strength	1.69e+010	N/m^2
Compressive Strength		N/m^2
Yield Strength	1.5e+010	N/m^2
Thermal Expansion Coefficient	10.7	/K
Thermal Conductivity	34	W/(m·K)
Specific Heat	461	J/(kg·K)
Material Damping Ratio		N/A

Figure 10.8

Assembly and Subassemblies

You should now have six parts modeled that will come together to make a bar puller. These parts are:

1. T Nut
2. Gripper
3. T Slot
4. Shank
5. Arm Adjust Nut
6. Arm

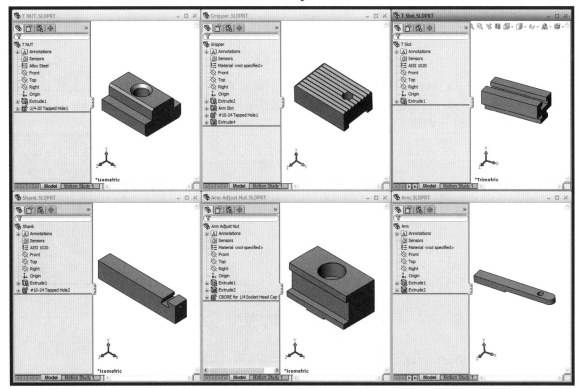

If they are not already open, open all six parts. With these parts, we will make one main assembly and one subassembly. One of the nice features of SolidWorks is that you can start an assembly and then realize that some of the parts should be separated into a subassembly. At that point, you can simply select the components of the new subassembly and tell SolidWorks to make them a subassembly. Or the subassemblies can be created separately, which we will do this time.

Assembly/Subassembly Creation

➢ With the **Arm Adjust Nut** open, select **File, Make Assembly from part**:

➢ Select your **ANSI-INCH** template, select **OK**

➢ In the **Begin Assembly** tab, under **Open Documents,** the part **Arm Adjust Nut** should be highlighted *(this is the part to be inserted into the assembly)*. If it is not highlighted, then select it. Select **OK**, and this part will be inserted at the assembly's origin:

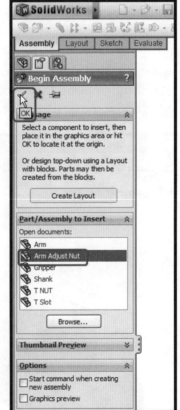

➢ Save this as **YourName_Arm Assembly**.

To see this being done, watch the video "**BP Sub Assembly**."

206 Chapter 10: Bar Puller

- Select **Insert Components**

- Select the **Arm** as the component to insert if it is open, otherwise **Browse** to the location where it is saved and open it.
- Move your mouse into the graphics area and click where you would like the arm to be located. Select somewhere that will place it near the other part but not over it.

The next step is to place constraints on this part so that it is positioned correctly. At this point, you should be able to click on the arm and drag it around the screen.

- Select **Insert, Mate** from the drop down menu or click on the paper clip icon

- Select the two faces as shown in Figure 10.9. The **Coincident** mate will be automatically selected. You can rotate your part around until you see the desired face of the arm, or you can right click on the face that is opposite it and then use the **Select Other** option and choose the desired face as shown in Figure 10.9.

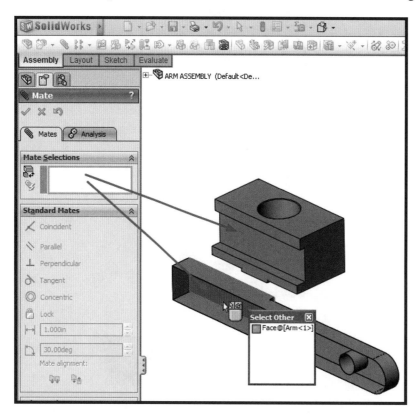

Figure 10.9

- Accept this selection, but **do not exit** the mate function.
- Again use the **Coincident** mate to mate the two surfaces as shown in Figure 10.10. Once the two surfaces are selected, right click to accept your selected mate.
- Lastly, mate the two end faces, coincident as shown in Figure 10.11 and accept these mates.

Figure 10.10

Figure 10.11

- **Save your work.**

To see this being done, watch the video "**BP Insert Component**."

Hole Series

The fasteners, clearance holes, and tapped holes for the 10-24 socket countersunk head cap screws can now be added all at once using the hole series wizard. This will ensure that all holes are in the correct alignment.

- Select the face of the **Arm** as shown in Figure 10.12.

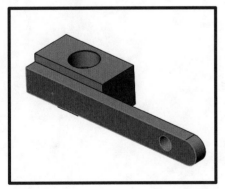

The hole series will create a clearance hole in our first part, a threaded hole in the second part, and add the fastener.

A hole series, unlike a hole wizard, starts with the sketch of the hole locations.

Figure 10.12

208   Chapter 10: Bar Puller

> From the **Assembly** tab, select **Assembly Features**, **Hole Series**. *(The point tool will be active.)*

> In the **Hole Position Properties Manager,** the **Create new hole** should be active. If it is not, select it

> Sketch two fully defined holes as shown in Figure 10.13.

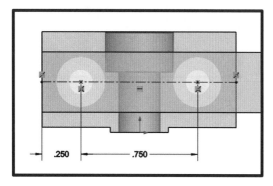

Figure 10.13

> Select the First Part tab
> Set up the **First hole** as shown in Figure 10.14. *(Your first hole has the six standard hole wizard clearance holes to choose from.)*

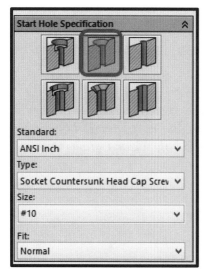

Figure 10.14

Chapter 10: Bar Puller   209

➢ Since there is no middle part, the **Middle Parts** section can be skipped. *(The middle part or parts can have either a standard drilled clearance hole or a straight slot.)*

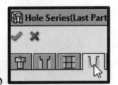

➢ Select the **Last Part** tab
➢ Select the **Arm Adjust Nut** as the **end Component** and set up this tab as shown in Figure 10.15.

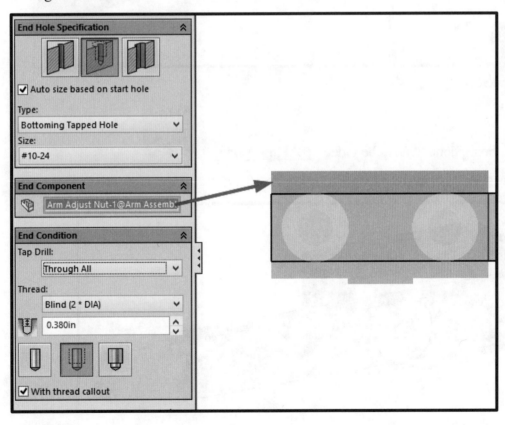

The end hole size is determined by the start hole. There are three hole type choices: tapped, clearance, or a clearance slot.

Figure 10.15

Chapter 10: Bar Puller

➢ Select the **Smart Fastener** tab and select all of the appropriate sizes and features as shown in Figure 10.16. Accept these selections.

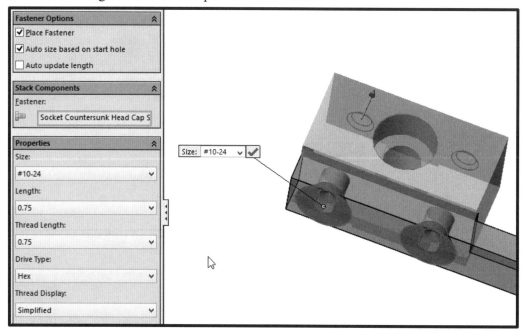

Figure 10.16

To see this being done, watch the video "**BP Hole Series**."

Your sub assembly should look similar to Figure 10.17.

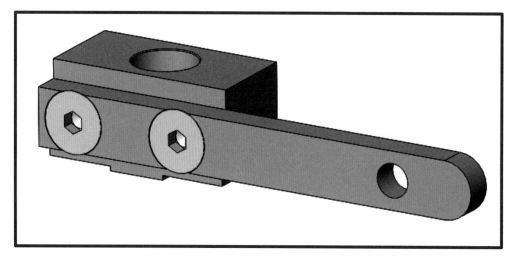

Figure 10.17

At the bottom of the feature tree, you will see the hole series and the smart fastener features

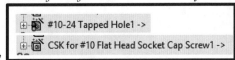

. *At the end of each part in the feature tree, you will see the hole wizard. The arrow at the end of the feature name tells you that this feature was created in the context of the assembly*

- Add the **Gripper** and fully mate it in the orientation as shown in Figure 10.18

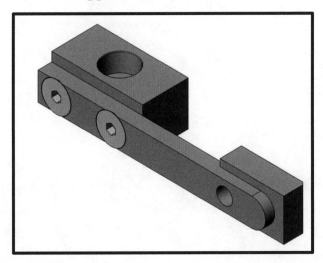

Figure 10.18.

- **Save your work.**

Adding Toolbox Components

- Select the **Design Library** from the right side of your screen:

- Select **Toolbox, ANSI Inch**

- Select **Bolts and Screws**, **Socket Head Screws**

- Click and hold down the mouse button on the **Socket Head Cap Screw** and drag this into the graphics area. If you are careful, you can release it in the correct position so that it will be properly constrained (**Coincident** and **Concentric** mates) once you release the mouse button as shown in Figure 10.19.

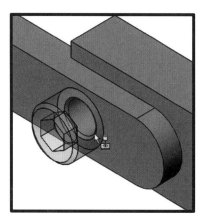

Figure 10.19

- Set up the screw to be a 10-24 with 3/8 thread length as shown in Figure 10.20.

Figure 10.20

- Accept this selection and press the **ESC** key to stop adding fasteners. *(You can continue to add fasteners until the ESC key is pressed.)*
- Constrain your screw if required.

To see this being done, watch the video "**BP Insert Fastener**."

Lock Rotation

- In the feature tree, expand the **Arm** and then expand the **Mates in Arm Assembly** folder. *(All of the mates to this part will be listed here, and all of the mates for the assembly will be in the bottom **MateGroup** folder .)*
- Find your **Concentric** mate to the **Socket Head Cap Screw** (SHCS), right click on this 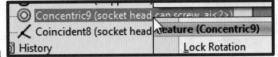 mate, and select **Lock Rotation**

Your screen should look similar to Figure 10.21.

Figure 10.21

- **Save your work** and close all parts.
- Start a new assembly document.
- Select your **ANSI-INCH** assembly template
- Insert your **Shank** into the assembly as the first component. *(Recall: the f beside the part name in the feature tree tells you that this first part in the assembly is fixed* 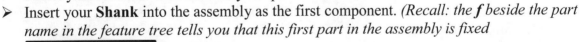 . *If you do not want it oriented as it currently is, you can right click on it and select **Float*** 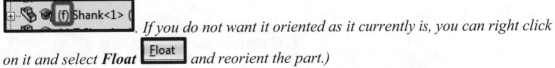 *and reorient the part.)*

To see this being done, watch the video "BP Assembly Main."

214 Chapter 10: Bar Puller

➤ Add the **T Slot** to the assembly by opening **File Explorer** , on the right side of your screen, navigating to your part, and then dragging and dropping it into the assembly.
➤ Mate the T slot to the shank so that it is centered on the shank.
➤ Add the **T Nut** to the assembly.
➤ Fully constrain the T Nut as shown in Figure 10.22.

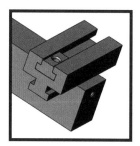

Figure 10.22

➤ Insert the **Arm Assembly**.
➤ Add appropriate constraints and an appropriate length 1/4-20 SHCS. Your screen should look similar to Figure 10.23.

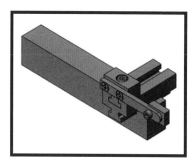

Figure 10.23

➤ **Save your work**.

Edit Toolbox Component

You may have inserted the wrong fastener or a fastener that was too short or your design has changed and therefore you are required to update your fastener. This can easily be done by editing the toolbox component.

➤ Right click on the 1/4-20 SHCS.
➤ Select **Edit Toolbox component** . This will bring you back to the original **Configure Component Property Manager**, where you can make any required changes.

➤ When you insert a fastener, you should notice the green arrow as shown in Figure 10.24. This shows you where the end of the fastener will be, based on the current length selected. You should also notice that you can set the thread length to less than the fastener length.

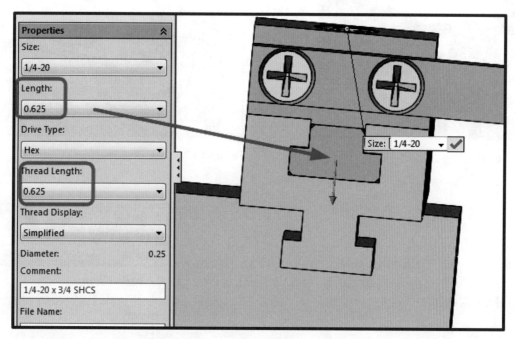

Figure 10.24

If you expand the subassembly in the feature manager tree, you might see feathers on some of the parts, as shown in Figure 10.25. This is SolidWorks saving memory and only loading these parts as lightweight. If you open one of these parts, it will then be resolved, or you can right click on an item and select **Set to Resolved**.

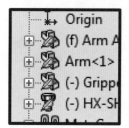

Figure 10.25

The remaining parts can now be copied from within the assembly.

➤ Hold the **CTRL** key and click on the **T Nut** *(in the graphics area or from the feature tree)*, drag it out away from the part, and release the mouse button to copy the part.

➤ **Save your work.**

To see this being done, watch the video "**BP Copy Parts**."

➢ Add a **Coincident** mate between the face of the T Nut to the T Slot as shown in Figure 10.26 and the **Coincident** mate as shown in Figure 10.27

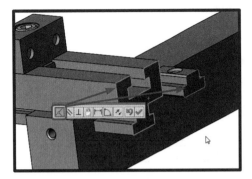

Figure 10.26

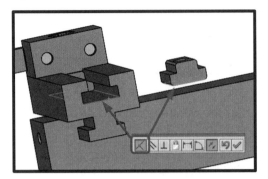

Figure 10.27

➢ Click on your **T Slot** and select **change Transparency**
(This will allow for you to easily mate this nut relative to the other T nut.)
➢ Add a **Coincident** mate between the two T Nuts as shown in Figure 10.28 and change the transparency back so that the part is no longer transparent

Figure 10.28

To see this being done, watch the video "**Change Transparency**."

➢ Mirror your **Arm Subassembly** so that your part now looks similar to Figure 10.29. *(Mirror through a plane in the center of the T Slot. If you do not have a plane there, edit the part and create a mid-plane.)*

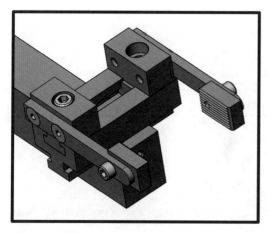

Figure 10.29

- Turn on **Create opposite hand version**
- Select **Next**
- Select the radio button to **Create new derived configuration in existing files**
- Select **Add Prefix**
- Select **Next**
- Check to **Transfer** Solid bodies and **Cosmetic threads** and accept these selections
- If you receive a message stating that components need to be resolved, select **OK**. *(If you have warnings in the mirrored component that will not resolve, then right click on the mirror and select **Dissolve Mirrored Component Features** and accept the recreation of mates in the top level assembly.)*

To see this being done, watch the video "**Mirror Subassembly to Create Opposite Hand Version**."

- Right click on your 1/4-20 SHCS and select **Copy with Mates**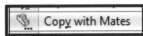
- Replace the two mates to the mirror component and accept these selections.
- Select the **Design Library**
- Select **Toolbox, ANSI Inch**
- Select **Set Screws (Socket)**
- Add a 10-24 x 1/2 cup point set screw  to lock the **Arm Adjust Nut** in place as shown in Figure 10.30.

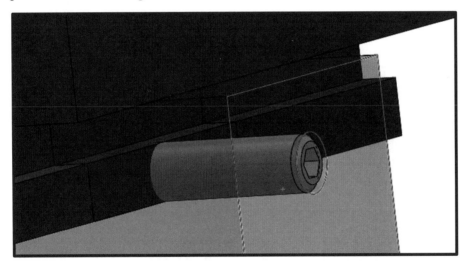

Figure 10.30

Exploded Views

- Open your **Arm Assembly.**
- Select the **Configuration Manager**
- Right click on your default configuration and select **New Exploded View:**

Chapter 10: Bar Puller **219**

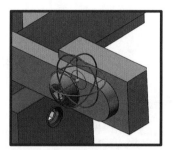

➤ Select the 10-24 SHCS

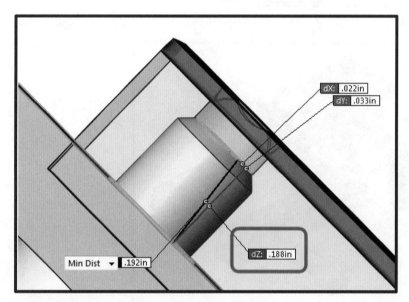

As shown in Figure 10.31, this fastener has 3/16 inch length of thread engagement. Since it has 24 threads per inch, it will required 4.5 turns to fully thread into the mating part.

This can be easily simulated in the explosion:
4.5 = (3/16) / (1/24)

Figure 10.31

➤ Select the blue circle as the rotary axis and the blue line as the explosion direction as shown in Figure 10.32.
➤ Add the **Settings** as shown in Figure 10.33 and select **Apply** and **Done.**

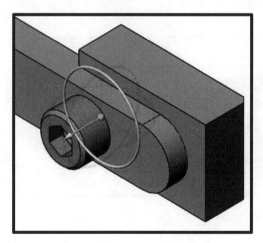

Figure 10.32

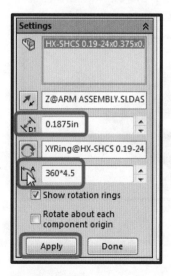

Figure 10.33

To see this being done, watch the video "**Threaded Explosion**."

- Select the fastener again and only the blue arrow.
- Set the distance to **0.5** and the **Explode Angle** to **0**
- Select **Apply** and **Done**
- Explode the rest of the components so that your explosion looks similar to Figure 10.34.

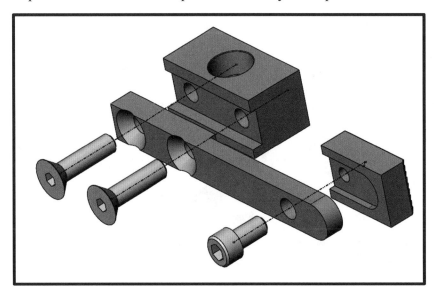

Figure 10.34

- **Save your work.** *(Be sure to include your exploded lines.)*
- Explode your main assembly to look similar to Figure 10.35. **Save your work.**
- When adding the exploded line sketch for the rectangular parts, select the edges that line up as shown in Figure 10.36.

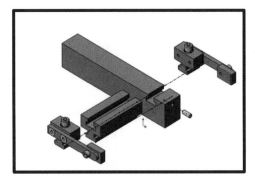

Figure 10.35

Figure 10.36

Animating Your Explosion

> ➢ In the **Configuration Manager,** collapse your explosion if it is in the exploded state by right clicking on the **ExpView1** and selecting **Collapse**

> ➢ Right click on the **ExpView1** and select **Animate explode**

The **Animation Controller** window will open as shown in Figure 10.37.

Figure 10.37

In section 1 of the controller, you have your standard play, pause, stop, rewind, fast forward, and stop buttons. In section 2, you can save an "AVI" file of the animation, and section 3 allows you to play the animation at regular speed (if neither the X1/2 nor the X2 button is depressed), half speed, or double speed. Section 3 also allows you to play the video from start to end once, to loop the playback, or to play from start to end and then play in reverse (collapsing).

> ➢ **Save and close your assembly**.

Motion Study

Motion studies can create very powerful animations and perform many calculations. We will simply use it here to create an animation of our assembly exploding.

> ➢ Open your **Arm Assembly.**

> ➢ From the lower left-hand corner, select the **Motion Study 1** tab

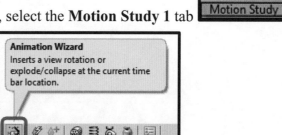

> ➢ Select the **Animation Wizard**

> ➢ Select the **Explode** radio button in the **Select an Animation Type** window:

Chapter 10: Bar Puller

- Select **Next**

- Set the **Duration** to **8.5** seconds and the **Start Time** to **0.5** and select **Finish**

- Select **Calculate**

- When the calculation is finished, select **Play**

- Again, you can save the animation

- Select the **Model** tab to go back to your model.

- **Save your work.**

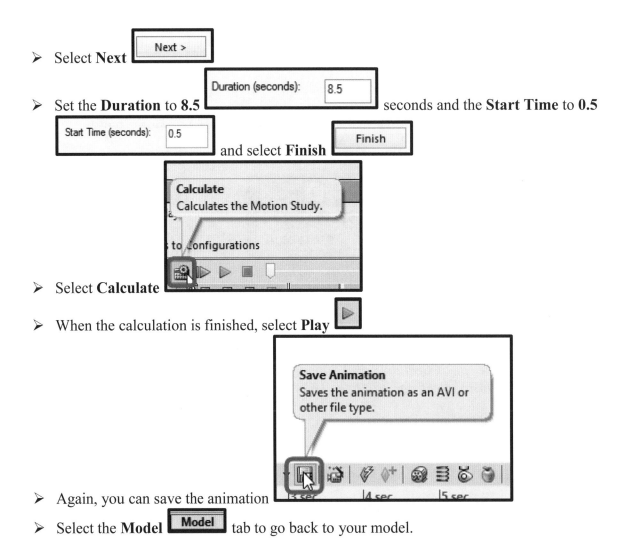

11. Bar Puller—Drawings

For your LEGO Minion, you created a drawing sheet set. For the Bar Puller, you will focus on creating individual drawing sheets and learning new drawing techniques.

- Open your bar puller shank.
- From the drop down menus, select **File**, **Make Drawing From Part**

- Select your **ANSI-B-INCH** size sheet
- Create the four views as shown in Figure 11.1.

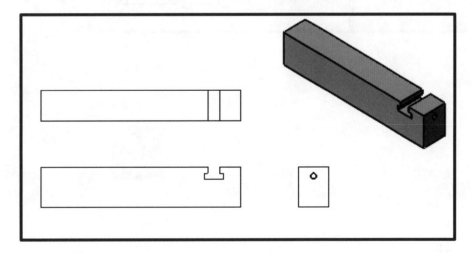

Figure 11.1

Break in View

- From the **View Layout** tab, select **Break**
- Select your front view.

223

224 Chapter 11: Bar Puller—Drawings

➢ Set up the **Broken View Properties** as shown in Figure 11.2. Click on the part in the front view approximately 1/4 of the way from the left side to place the first break line and then click approximately 1/4 of the way on the part from the right side to place the second break line and complete the break view.

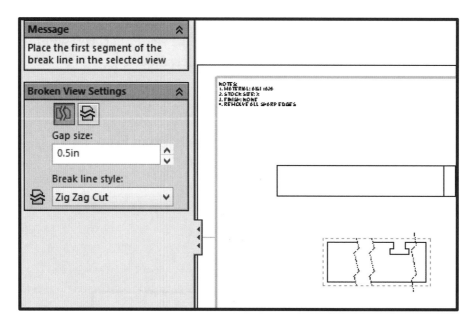

Figure 11.2

➢ Accept these selections.

To see this being done, watch the video "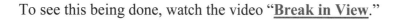Break in View."

Crop View

➢ Sketch a rectangle in the top view as shown in Figure 11.3.

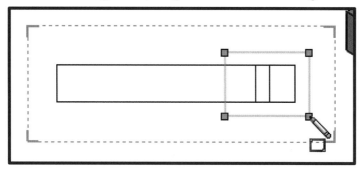

Figure 11.3

- ➢ With the rectangle selected, choose **Crop View** from the **View Layout** tab.
- ➢ Your top view should now look similar to Figure 11.4.

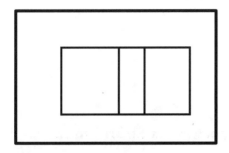

Figure 11.4

To see this being done, watch the video "**Creating a Cropped View**."

Hole Callout

- ➢ From the **Annotation** 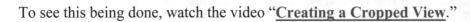 tab, select **Hole Callout**
- ➢ Select the tapped hole in the right side view.
- ➢ Move your mouse in-between the front and right side views and click to place the hole callout as shown in Figure 11.5.

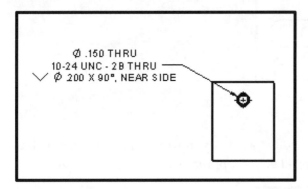

Figure 11.5

To see this being done, watch the video "**Creating a Hole Callout**."

- ➢ **Save your work** and close the drawing.

Ordinate Dimensioning

➤ On an inch ANSI A size sheet, create a drawing with the top view of your **Arm** as shown in Figure 11.6.

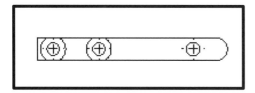

Figure 11.6

➤ From the **Annotation** tab, select the drop down arrow below **Smart Dimension**  and then select **Horizontal Ordinate Dimension**:

➤ Select the far left edge of the part as the zero for the ordinate dimension as shown in Figure 11.7.

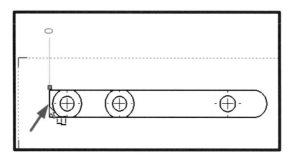

Figure 11.7

➤ Working from left to right, select each feature that you would like dimensioned, until your drawing looks similar to Figure 11.8. Accept these selections.

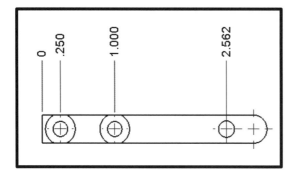

Figure 11.8

➢ Right click on any of the ordinate dimensions and select **Add To Ordinate:**

➢ Add the two dimensions as shown in Figure 11.9. *(Recall: to dimension to the outside edge of a circle, you must hole the **Shift** key.)*

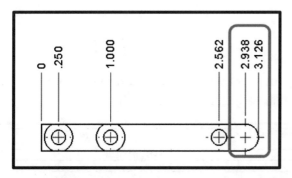

Figure 11.9

➢ Exit ordinate dimensioning.

To see this being done, watch the video "**Creating Ordinate Dimensions**."

Hole Tables

➢ From the **Annotations** tab, select **Tables, Hole Table**
➢ In the **Hole Table Properties**, the **Datum, Origin** section should be highlighted blue. If it is not, click in the field
➢ Select the lower left-hand corner of the part as the origin as shown in Figure 11.10.

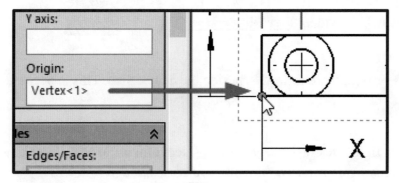

Figure 11.10

228 Chapter 11: Bar Puller—Drawings

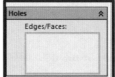

➢ The **Holes** field should now be highlighted blue. If it is not, click in the field.

➢ Select the edge of the hole circles and right click when done
➢ Accept these selections and click on your sheet where you would like the hole table placed. Your drawing should look similar to Figure 11.11.

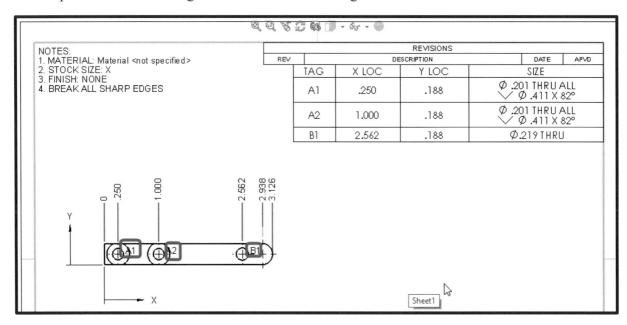

Figure 11.11

➢ The hole labels, **A1** and **A2,** are hard to read. To move them, click and hold on the label, then drag and drop it to the desired location to neaten up your drawing.

To see this being done, watch the video "<u>Creating a Hole Table</u>."

Stock Size

In Figure 11.11, in the notes section, there is an "**X**" beside the **Stock Size**. This is because the part property is left blank.

➢ In your drawing sheet, right click near the view and select **Open Part** . *(The view will have an orange rectangle around it once you are close enough to select it.)*
➢ In the part, from the drop down menus, select **File** , **Properties**

- Select the **Custom** [Custom] tab.
- Set the **Stock Size** to **1/4 x 3/8 x 3 1/4**:

StockSize	Text	1/4 x 3/8 x 3 1/4

- Accept these selections and rebuild, save, and close the part.
- The stock size should now appear in the notes section of your drawing:

  2. STOCK SIZE: 1/4 x 3/8 x 3 1/4

12: Centering Jig—Parts

A centering jig can be used to ensure that all parts within its clamping range are always held with their center in the same location. This jig uses a left- and right-hand thread to move the jaws symmetrically about the center. The spring helps eliminate any backlash inherent with the thread. A tight tolerance design intent is critical, as the position and fits will determine accuracy of all parts held in the jig.

This particular jig, as shown in Figure 12.1, was designed for a laser engraver. All parts for this jig are **INCH.**

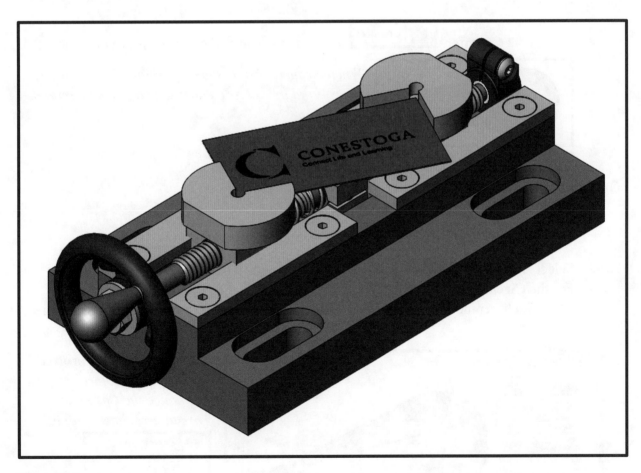

Figure 12.1

Handle Grip

- Create a revolved part as shown in Figure 12.2.

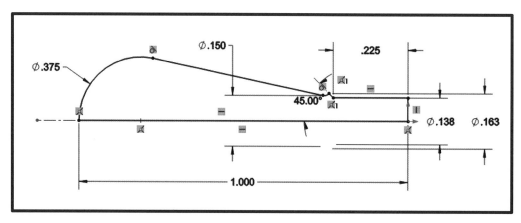

Figure 12.2

- Add a **0.02 Chamfer** 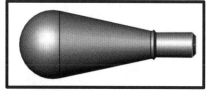 feature to the end of the part as shown in Figure 12.3.

When a round part as shown in Figure 12.3 requires an external thread, you can create a cosmetic thread.

Figure 12.3

Cosmetic Thread

- From the **Insert** drop down menu.
- Select **Annotations** and then **Cosmetic Thread**
- For the **Circular Edge,** select the edge as shown in Figure 12.4, set the **Standard** to **ANSI Inch**, set the **Size** to **6-32**, and set the **End Condition** to **Up to Next.** Accept these selections.

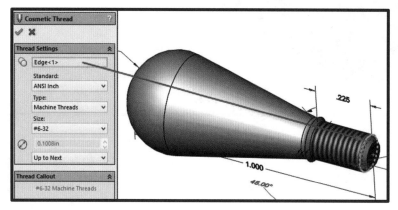

*If you expand your revolved feature in the feature tree, you will notice the **Cosmetic Thread** associated with it:*

Figure 12.4

Chapter 12: Centering Jig—Parts   233

To see how to create cosmetic threads, watch the video "<u>Creating Cosmetic Threads</u>."

➢ Set the material to **AISI 1020 Steel, Cold Rolled**

➢ **Save your work** as Grip_YourName.

Handle

In your right plane, create a revolved part from the sketch shown in Figure 12.5.

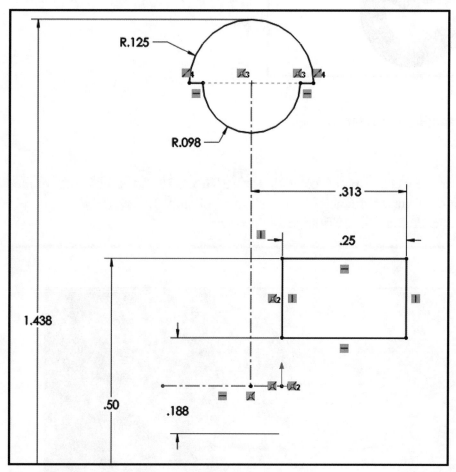

Figure 12.5

Your feature tree will show this as two separate **Solid Bodies**

Do not worry about this, as these will be joined into one solid body shortly.

Creating a Basic Loft

➢ Create two planes offset upward from the top plane, one **0.225** and one **0.515** as shown in Figure 12.6.

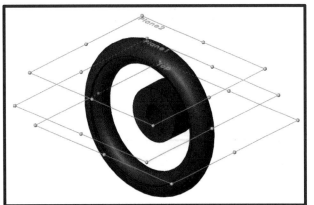

Name the 0.225 offset plane *T225* and the 0.515 offset plane *T515*.

These planes will be used to create sketches for a lofted feature. This will be the first feature that you create which cannot be started by selecting the feature but requires at least two sketches.

Figure 12.6

➢ **Save** your part as **Handle_YourName**.

➢ In the **T225** plane, create a center point rectangle 3/16 x 1/16. Lock the center of the rectangle to the center of the 0.5 diameter hub as shown in Figure 12.7. After the rectangle is fully defined, add a 0.025 fillet to each corner.

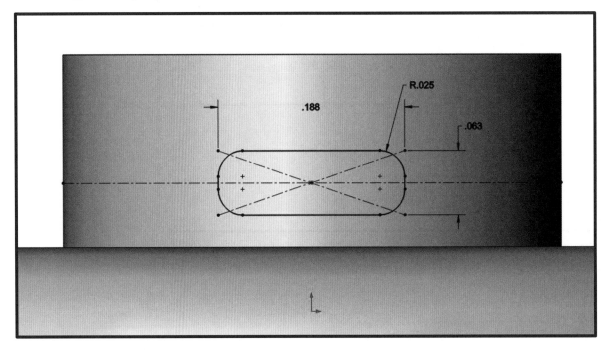

Figure 12.7

➢ In the **T515** plane, create another center point rectangle as shown in Figure 12.8.

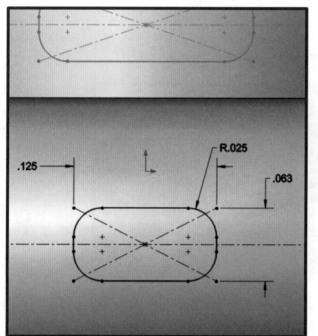

Lock the center of this rectangle to the center of the 1 7/16 diameter hub. The center line shown in both Figures 12.7 and 12.8, will be coincident with each edge of the hub:

Figure 12.8

➢ In the right plane, create the sketch as shown in Figure 12.9. Name this sketch **CLSketch**.

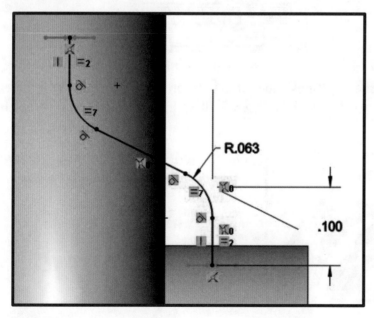

*The lower endpoint of the line will be coincident with the sketch in the **T225** plane. The upper endpoint of the line will be coincident with the sketch in the **T515** plane.*

Figure 12.9

➢ Create the three lines first, fully define them, and then add the fillets.

236 Chapter 12: Centering Jig—Parts

- From the **Features** tab select **Lofted Boss/Base**
- In the **Loft Property Manager**, for the **Profiles,** select the sketch in the **T225** plane and then select the sketch in the **T515** plane as shown in Figure 12.10.
- *It is **critical** that you select each sketch at the same approximate location.* Select each sketch just to the right of the green dots as shown in Figure 12.10.

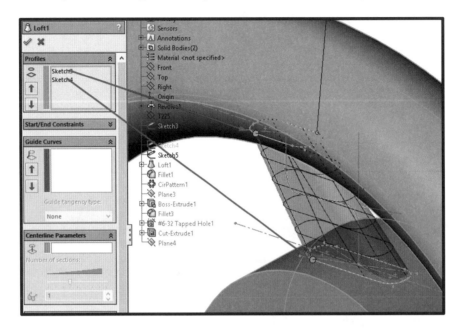

Figure 12.10

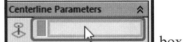

- Click in the **Centerline Parameters** box.
- Select the **CLSketch** sketch as the center line path for the loft to follow.
- Accept the remaining default settings and exit the loft property manager. Your part should now look similar to Figure 12.11. There will no longer be two solid bodies listed in the feature tree.

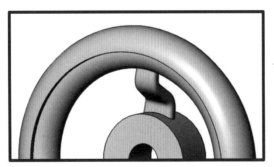

Figure 12.11

A loft feature joins two or more different sketch profiles into a smooth transitioning feature.

To see this basic loft being created, watch the video "**Creating a Basic Loft**."

- Add a 1/32 feature fillet to each end of the loft as shown in Figure 12.12.
- Create a circular feature pattern of the loft and the fillet as shown in Figure 12.13

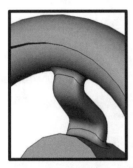

Figure 12.12

Figure 12.13

On Edge Plane

- Select the edge of your part as shown in Figure 12.14.

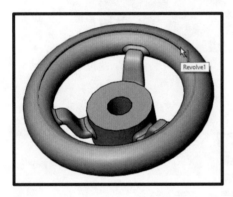

Figure 12.14

- Select **Reference Geometry**, **Plane**
- Accept the creation of the plane as shown in Figure 12.15. Name this plane **For Thread**.

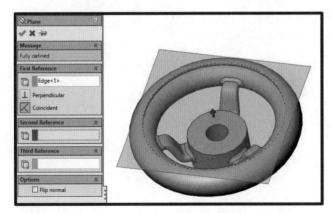

Figure 12.15

➢ Create a sketch with two three point arcs in this plane as shown in Figure 12.16.

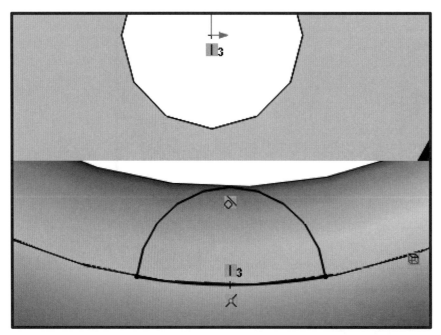

You can use the convert entities command to quickly create the larger diameter 3 point arc.

Figure 12.16

➢ Exit the sketch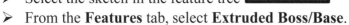

➢ Select the sketch in the feature tree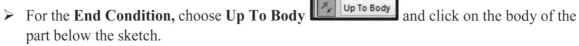
➢ From the **Features** tab, select **Extruded Boss/Base**.

➢ For the **End Condition,** choose **Up To Body** and click on the body of the part below the sketch.
➢ Accept this selection and exit the extrusion.

To see this being done, watch the video "**Extrude up to Body**."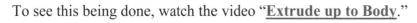

Chapter 12: Centering Jig—Parts 239

Hole Wizard

The hole wizard allows for the quick and easy creation of one or more tapped or clearance holes. It combines two sketches into one feature. One sketch contains all of the hole locations, and the other sketch contains the cross sectional profile of the hole.

> Select the **For Thread** plane and look **Normal To** it (this means to look perpendicular to the selected face or surface.)

> With the plane selected, from the **Features** tab, select **Hole Wizard**

> From the **Hole Specification Property Manager**, in the **Hole Type** section, 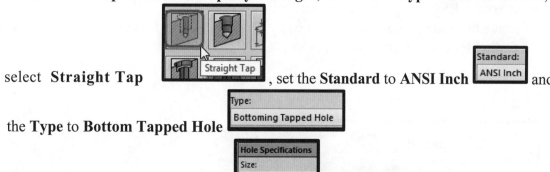 select **Straight Tap**, set the **Standard** to **ANSI Inch** and the **Type** to **Bottom Tapped Hole**

> Set the **Hole Specifications** to **6-32**

> Set the **End Condition** to **Through All** 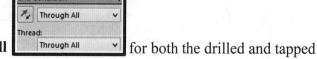 for both the drilled and tapped holes.

> Set the **Options** as shown in Figure 12.17.

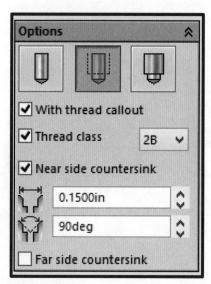

Figure 12.17

You have now set all of the conditions for the hole type, these will determine the cross-sectional profile for the revolve cut that the hole wizard will make at the desired location. Next you will set the location of the hole(s).

➢ Select the **Position** tab. If asked to select the face for the hole or slot position, then select the **For Thread** plane from the feature tree. *(You should notice the **Point** icon is now active.)*
➢ Click on the part edge you used to create the plane, roughly opposite the first loft that you created.
➢ Press the **ESC** key to stop adding holes.
➢ Make the point position fully defined by adding a **Vertical** constraint to the origin and a **Coincident** constraint to the circular edge.
➢ Exit the hole wizard.

Your part should now look similar to Figure 12.18.

Figure 12.18

➢ **Save your work.**

Hole Specification Property Manager

In Section 1, of the **Hole Specification Property Manager,** you can select the type of hole. As shown in Figure 12.19, there are nine types of holes that can be selected:

Type A will create a counterbored hole.

Type B will create a countersunk hole.

Type C will create a drilled hole.

Type D will create a straight tapped hole.

Type E will create a tapered tapped hole.

Type F will create a legacy hole (these are user-defined holes).

Type H will create a counterbored slot.

Type I will create a countersunk slot hole.

Type J will create a straight slotted hole.

In Section **2**, you set the hole specifications, such as the tap type and size or the drill diameter or the type of fastener the clearance hole is used for.

In Section **3**, you set the end condition for the hole, such as through the entire part or to a specified depth in the case of a blind hole.

In Section **4**, you set options for the hole, such as countersinks on various locations of the hole.

Figure 12.19

After setting the hole specifications, the **Positions** tab is selected, and the hole locations will be added to create the desired holes in the part.

For a detailed explanation of the hole wizard, watch the video "<u>Hole Wizard</u>."

242 Chapter 12: Centering Jig—Parts

> Select the face of the wheel as shown in Figure 12.20 and look **Normal To** it.

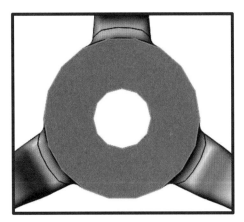

Figure 12.20

> With the face selected, choose **Extruded Cut** from the **Features** tab.
> Create the sketch of the three lines, one 3 point arc *(or converted edge)*, and the center line as shown in Figure 12.21.

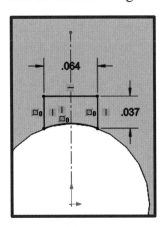

Figure 12.21

Bilateral Dimension

> Select the **0.037** dimension and change it to **0.0372.**

> Set the **Units Precision** to **4** decimal places

> In the **Dimension Properties, Tolerance/Precision** section, select **Bilateral** as the dimension type.

> Enter 0.005 in the plus section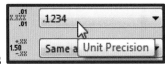

Chapter 12: Centering Jig—Parts **243**

- Accept these selections ✓
- Create a bilateral tolerance dimension for the **0.064** dimension with the plus dimension being **0.002**.

Your sketch should look like Figure 12.22.

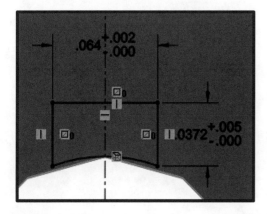

Figure 12.22

- Exit the sketch
- Set the extrusion end condition to **Up To Next** [Up To Next] and accept these selections.

Your part should look similar to Figure 12.23.

Figure 12.23

➢ Add a 1/64 fillet to the part edges as shown in Figure 12.24.

This handle represents a cast part and casting will often have manufacturing marks cast into them. These and the cast appearance will be added next.

Figure 12.24

➢ Set the material as **Gray Cast Iron** [Gray Cast Iron]

➢ **Save your work**.

Adding Raised Text

Raised text is often used on cast parts as a means to identify the time, location, and pattern used to create the part. Should a problem arise later, this aids in analyzing the root cause and in preventing repetition of the same problem.

Angled Plane Creation

➢ Select your right plane in the feature tree [Right Plane]

➢ Select **Reference Geometry, Plane** [Plane]

➢ In the **Plane Property Manager** select **At Angle** and enter **30** [30.00deg]

➢ From the drop down menus across the top, select **View** [View] **Temporary Axes:** [Temporary Axes]

- For the **Second Reference,** select the temporary axis through the part center:

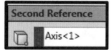

- If your axis is not as shown in Figure 12.25, then **Flip** it

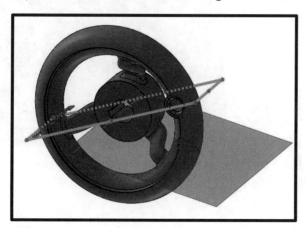

Figure 12.25

- Accept these settings.
- Name this plane **30 to Right** and turn off your temporary axis.
- Create an offset plane of the **30 to right** plane, which is offset **0.75** as shown in Figure 12.26.
- Call this plane **30 Offset** *(Be certain that this plane is opposite the loft as shown in Figure 12.26.)*

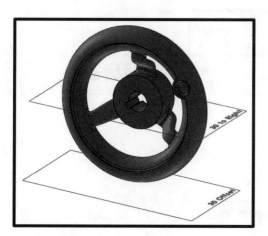

Figure 12.26

- Look **Normal To** your **30 Offset** plane *(If when you look **Normal To** the plane and you cannot see all three lofts, then select look **Normal To** again.)*

➢ Click on the **30 to Offset** plane in the feature tree and select **Sketch**
➢ Add a center line as shown in Figure 12.27.
➢ Offset this center line **1/32** to the left.

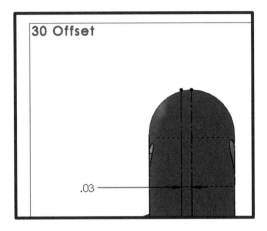

Figure 12.27

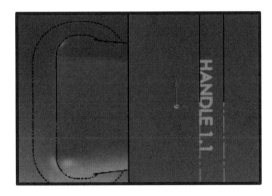

Figure 12.28

➢ Select the line. In the **Line Properties, Option** section, turn on **For construction** ☑ For construction to change this to a center line.

➢ From the **Sketch** menu, choose **Text** . *(If you accidentally leave the text property, simply double click on the text to return.)*

➢ Select the offset center line as the **Curves** Curves to create the text along it. *(The field will be active as soon as you select the text option.)*

➢ In the **Text** field, enter **HANDLE 1.1** [HANDLE 1.1]. *(If your text does not look like that shown in Figure 12.28, then you may need to **Flip Horizontal** AB|BA or **Flip Vertical** A|∀ .)*

➢ **Center Align** the text

➢ Uncheck **Use document font** ☐ Use document font

➢ Select **Font** [Font...]

Chapter 12: Centering Jig—Parts **247**

➢ Set the font as shown in Figure 12.29 and select **OK.**

Figure 12.29

➢ Exit the **Text Property Manager**. *(The sketch will not be fully defined.)*

When you exit the Text Property Manager, there will be a point created on the partorigin that is blue in color. Only this point needs to be constrained or dimensioned in order for the text to become fully defined. The text will always be blue in color.

➢ Click on the blue point created with the text, it will be located at the origin. Drag it off and then back onto the origin to add the **Coincident** mate and fully define the text.
➢ Exit the sketch

➢ Select the sketch in the feature tree
➢ Select **Extruded Boss/Base**.

➢ Set the extrusion distance to **0.005**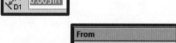

➢ In the **From** field, select **Surface/Face/Plane**
➢ Again in the **From** field. the **Select A Surface/Face/Plane** selection box should be active (blue in color). Select the face as shown in Figure 12.30.

If the text is not as shown in Figure 12.30, then reverse the direction of the extrusion

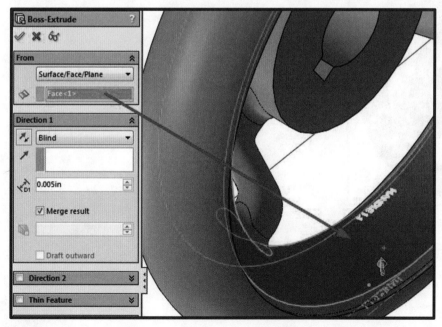

Figure 12.30

248   Chapter 12: Centering Jig—Parts

- Accept these selections.
- **Save your work**.

To see this being done, watch the video "**Adding Raised Text**."

Adding an Appearance

- At the top of the feature tree, right click on your part name.
- Select the down arrow beside the beach ball *(Selecting the beach ball icon allows you to set the appearance of features.)*

- Select the **Handle**
- On the right side of your screen, expand the **Metal** appearances
- Select **Iron**

- In the **Default Appearance color** section, select **Cast Iron**

Your part should now look similar to Figure 12.31.

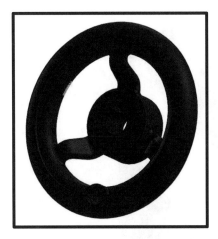

Figure 12.31

- Accept these selections.
- **Save your work**.

To see this being done, watch the video "**Adding a Cast Iron Appearance**."

Bushings

- Create the bushing as shown in Figure 12.32.
- Make the material **Tin Bearing Bronze** (*a Copper Alloy*).
- **Save** this part as **Bushing_YourName**.

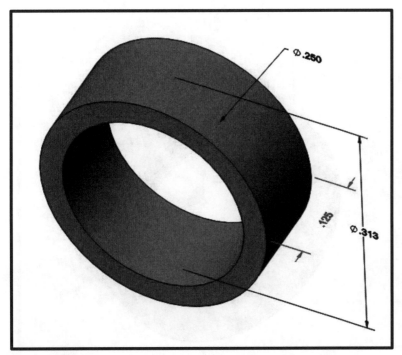

Bushings often require a small clearance on their ID and a small interference on their OD.

To add this, you will use a standard Fit tolerance. The exact fit is determined based on function and feature sizes.

There are three ways to display a Fit tolerance. To minimize error and decrease production time, I would always encourage you to use the Fit with tolerance option.

Figure 12.32

Fit with Tolerance Dimension

- Double click on your part so that the dimensions are displayed as shown in Figure 12.32.
- Select the extension line of the **0.25** dimension to open the **Dimension Property Manager**.
- In the **Tolerance/Precision** field, select **Fit with Tolerance**
- In the **Classification** field, select **Clearance**
- In the **Hole Fit** field, select **H7**
- Enable the **Show parentheses** option
- Set the **Unit Precision** to **4** decimal places
- Set the **Tolerance Precision** to **Same as nominal**

- ➤ Accept these selections
- ➤ Set the **0.313 diameter dimension** as shown in Figure 12.33.

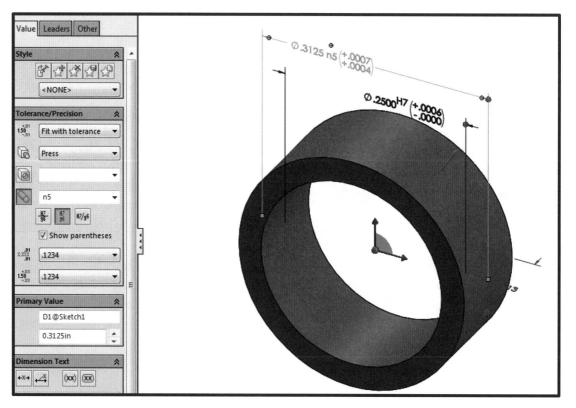

Figure 12.33

- ➤ **Save your work**.

Stock Size

- ➤ From the drop down menu, select **File, Properties**
- ➤ Select the **Custom** tab
- ➤ For the **Stock Size** 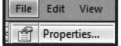 property, enter **1/2**.
- ➤ Select **OK**.

If the property is missing, add it by typing the property name or by selecting it from the drop down menu. By only entering the 1/2 as the stock size, you are telling the manufacturer to use any length that will work.

*Open your **Grip** part and make its **Stock Size 1/2 x 1 1/8**. We enter stock size as fractions, because the individual who will make this part will typically check the stock using a measuring tape or pocket ruler.*

Screw

> Create a revolved part in the front plane as per Figure 12.34. **Note the location of the part origin.**

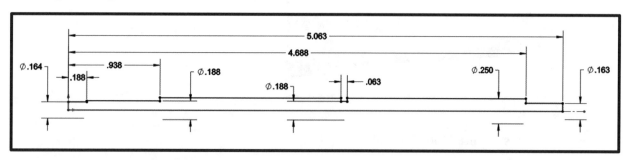

Figure 12.34

> Add a **0.016 x 45°** chamfer to each end.
> Add a **1/32 x 45°** chamfer to the two locations as shown in Figure 12.35.

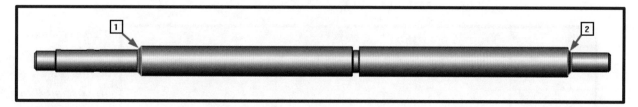

Figure 12.35

> Add a **0.008** fillet to the five locations as shown in Figure 12.36.

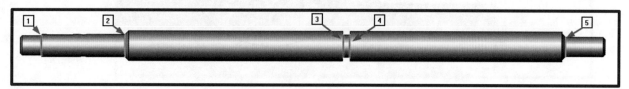

Figure 12.36

> **Save** your part as **YourName_Screw**.
> Set the material to **AISI 1020 Steel, Cold Rolled**.

Through All Both

> Select the **Top** plane in the feature tree 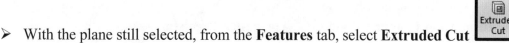 and look **Normal To** it.

> With the plane still selected, from the **Features** tab, select **Extruded Cut**.
> Sketch a **Ø1/16 hole** on center and **0.546** from the origin.

> Exit the sketch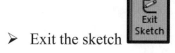

252   Chapter 12: Centering Jig—Parts

- For the **End Condition,** choose **Through All - Both**
- Accept these settings, and your part should now look similar to Figure 12.37.

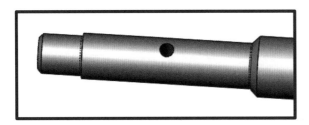

Figure 12.37

- Name this feature **Roll Pin Hole**.

- Select the **Front** plane from the feature tree, look **Normal To** it.
- From the **Features** tab, select **Extruded Cut**.
- Create the sketch as shown in Figure 12.38.

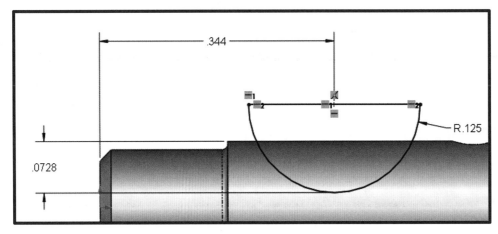

Remember to hold shift while selecting the arc so that you can add the dimension as shown.

Figure 12.38

Diameter Dimension

- Right click on the **R0.125** dimension.
- Select **Display Options, Display As Diameter**

Limit Dimension

- Select the **Ø0.250** dimension.
- In the **Dimension Property Manager**, **Tolerance/Precision** section, select **Limit**  as the **Tolerance Type.**
- In the **Maximum Variation** field, enter **0.003** and accept these settings.
- Exit the sketch.
- Make the **Extruded Cut** a mid-plane extrusion 1/16 inches wide.

➤ Rename this feature **#202 Woodruff**
➤ Double click on the feature so that the dimensions appear and edit the **.0625** dimension to a limit dimension of **0.063/0.0615** and the **0.0728** dimension to a bilateral dimension with a **+0.005** tolerance as shown in Figure 12.39.

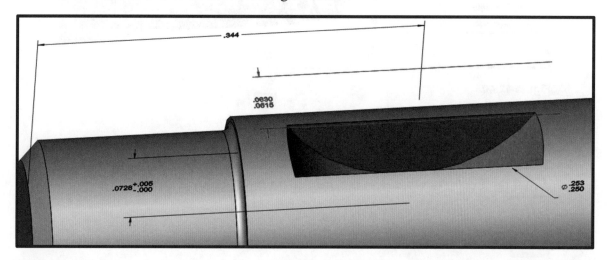

Figure 12.39

➤ Add a #8-32 cosmetic thread to the Ø0.164 x 3/16 long end of the shaft:

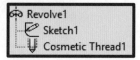

Sweep Cut Thread

If this jig was to be created on a rapid prototype machine, such as a 3D printer, then the cosmetic threads you just added would not be actual threads, but just the cylindrical features. To 3D print an actual thread, you will need to model it using a sweep cut feature.

➤ Create a plane in the middle of the **1/16 inch** wide groove using the mid-plane option.
➤ Name this plane **Mid Plane**.
➤ From this plane, create an offset plane **2 inches** towards the #202 Woodruff key feature.

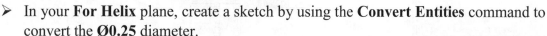

➤ Name this plane **For Helix**
➤ In your **For Helix** plane, create a sketch by using the **Convert Entities** command to convert the **Ø0.25** diameter.

➢ Exit the sketch. Your screen should look similar to Figure 12.40.

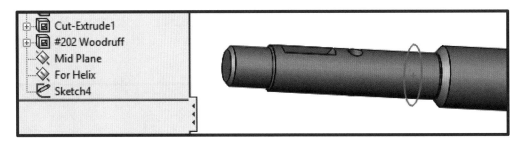

Figure 12.40

Helix

➢ Select this sketch in the feature tree
➢ From the **Features** tab, select the drop down arrow below **Curves**:

➢ Select **Helix and Spiral**
➢ Set up the **Helix/Spiral Property Manager** as shown in Figure 12.41. Accept these selections.

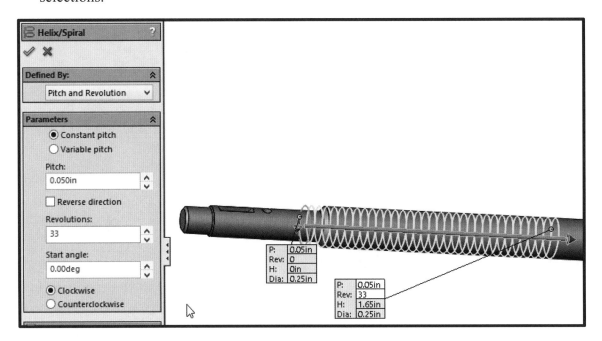

Figure 12.41

To see this being done, watch the video "<u>Creating a Helix</u>."

Creating the Thread Profile

- Select and look **Normal To** the top plane.
- Select the top plane again and select [Top] **Sketch**. *(Recall: this is one way to start a sketch on a plane.)*
- Create the sketch as shown in Figure 12.42.

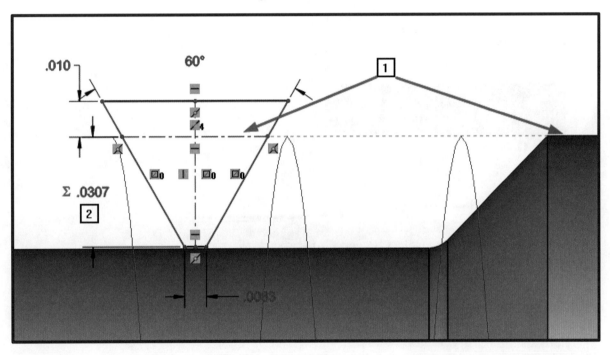

Figure 12.42

The dimensions are purple because the **Mark For Drawing** option has been unchecked by right clicking on the dimension. The horizontal center line and the part edge are collinear. The equation for the single depth of thread is: $= 0.61343 \times (1/20)$

- Right click on the horizontal center line and choose **Select Midpoint** as shown in Figure 12.43.

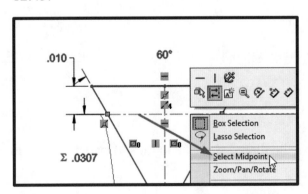

Figure 12.43

- Hold the CTRL key and select the helix near the end as shown in Figure 12.44. Add the **Pierce** relationship to fully define your sketch. *(Alternatively, you could make this midpoint coincident with the end of the spiral.)*

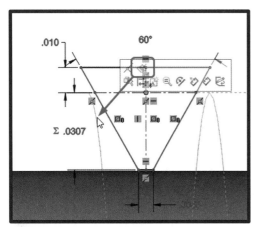

Figure 12.44

- Exit the sketch

- From the **Features** tab, select **Sweep Cut**
- In the **Cut-Sweep Properties Manager** and in the **Profile and Path** section, select the thread profile sketch that you just created as the **Profile** and then select the helix as the **Path**
- Accept these selections, and your part should look similar to Figure 12.45.

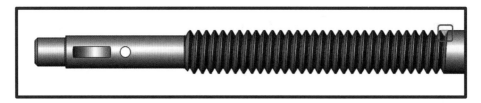

Figure 12.45

- **Save your work.**

To see this thread profile and sweep cut being created, watch the video "**Thread-Swept Cut**."

To see an alternative method for creating the thread sketch and the cut sweep, watch the video "<u>Sweep Cut Thread</u>."

Revolve Cut

> Select the face, shown in blue in Figure 12.45, where the sweep cut ended.

> Select **Revolve Cut** from the **Features** tab.
> With the face still selected, choose **Convert Entities.** *(You should now have a fully defined sketch of the thread profile.)*
> Exit the sketch.

> From the **View (Heads-Up), Hide/Show Items** drop down menu, select 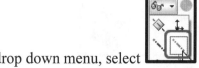 **Temporary Axes.**
> Select the **Temporary Axes** in the middle of the shaft as the **Axis of Revolution**:

> Accept these selections. Your part should look similar to Figure 12.46.

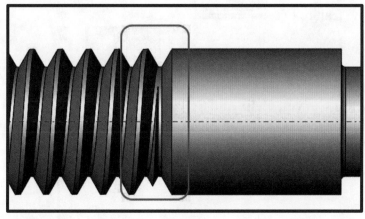

Figure 12.46

> Turn off the display of the **Temporary Axes**.
> Create a **Midplane** in the center of the **0.063** wide groove.
> Mirror the cut sweep and the revolve cut through this plane. Your part should now look similar to Figure 12.47.

Figure 12.47

> **Save your work.**

Spring

The spring shown in Figure 12.48 will be created using a composite curve and a swept boss. A composite curve is made by joining multiple sketches into one curve. This is required because the swept boss feature requires one path sketch and one profile sketch.

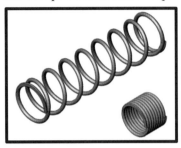

Figure 12.48

Swept Boss

➢ In the front plane, create a fully defined Ø**0.329** circle whose center is locked to the origin.

➢ Select your sketch from the feature tree

➢ From the **Features** tab, select **Curves, Helix and Spiral**
➢ Set up the **Helix/Spiral** parameters as shown in Figure 12.49. Accept your selections.

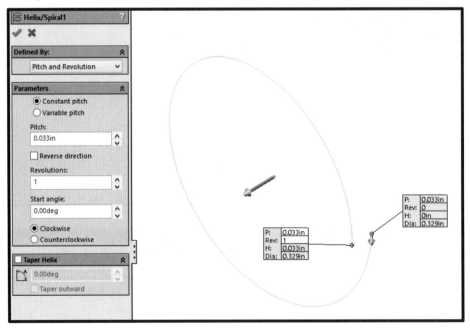

Figure 12.49

Derived Sketch

- In the feature tree, expand the **Helix/Spiral**.

- Select the **Sketch** , hold the **CTRL** key, and select the **Front** plane:

- From the drop down menu, select **Insert** **Derived Sketch**

To see this being done, watch the video "**Derived Sketch**."

A derived sketch is an exact duplicate of the copied sketch in the new plane. It cannot be dimensioned. Therefore, it must be fully defined using relationships. Your feature tree will also show it as a derived sketch

- From this derived sketch, create another **Helix/Spiral** using the conditions as shown in Figure 12.50. Your screen should look similar to Figure 12.51.

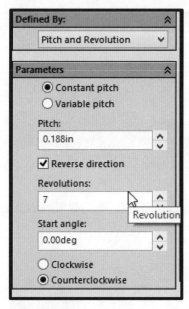

Figure 12.50

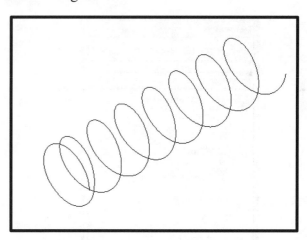

Figure 12.51

260 Chapter 12: Centering Jig—Parts

➢ Create a new plane parallel to the **Front** plane, where the **Second Reference** is the end point of the second helix as shown in Figure 12.52.

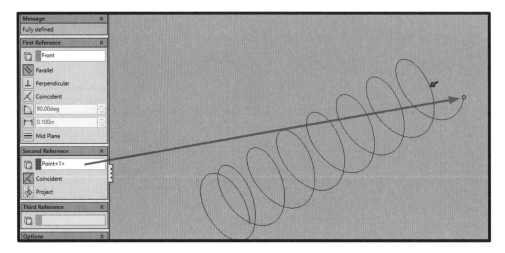

Figure 12.52

➢ Create another fully defined **derived sketch** of the initial sketch in this plane. *(Use a coincident relationship between the circle center and the origin.)*

➢ Using this latest derived sketch, create another **Helix/Spiral** with the conditions shown in Figure 12.53. Your screen should now look similar to Figure 12.54.

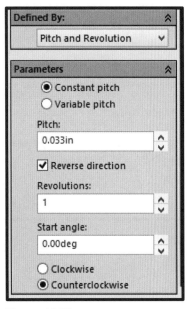

Figure 12.53

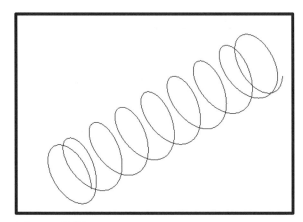

Figure 12.54

Composite Curve

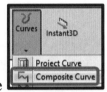

- From the **Features** tab, select **Curves, Composite Curve**
- Select your three **Helix/Spiral** from the feature tree or from the graphics area as the entities to join . Accept these selections.
- **Save your work** as **YourName_Spring**.

You should notice in your feature tree that these three spirals have been consumed by the composite curve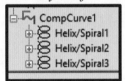

To see this being done, watch the video "**Composite Curve**."

- In the top plane, sketch the fully defined circle as shown in Figure 12.55.

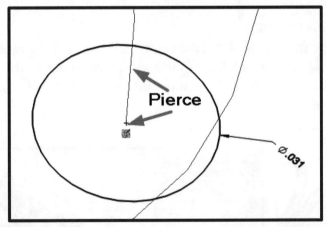

Figure 12.55

- Name this sketch **Profile**.

- From the **Features** tab, select **Swept Boss/Base**
- Select the sketch you just created as the **Profile**. *(This can be selected from the feature tree or from the graphics area. Your sketch numbers do not need to be the same as the ones shown.)*
- Select the composite curve as the **Path**
- Accept these selections. Your part should now look similar to Figure 12.56.

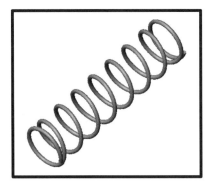

Figure 12.56

To see this being done, watch the video "Sweep Boss."

- **Save your work.**

- Show your **Profile** sketch
- Create a new plane parallel to the front plane using the center point of the **Profile** sketch as the **Second Reference** as shown in Figure 12.57. Name this plane **Front End**.
- Create an offset plane that is offset **1.375** from the **Front End** plane as shown in Figure 12.58. Name this plane **Back End**.

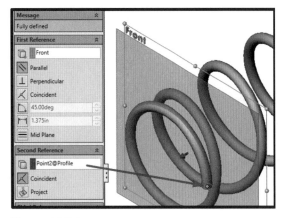

Figure 12.57

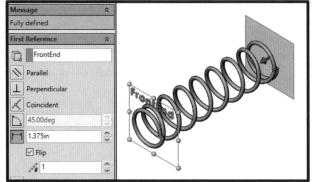

Figure 12.58

Extruded Cut with a Line

➢ In the top plane, sketch the line as shown in Figure 12.59. Only use sketch relations to fully define the line.

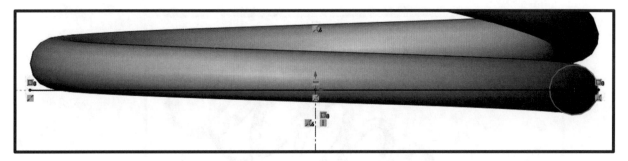

Figure 12.59

➢ Exit the sketch.

➢ Select **Extruded Cut** . *(Recall: the sketch should be selected in the feature tree before selecting the **Extruded Cut**. If it was not preselected, then select it when prompted to select a sketch.)*

➢ Set up the **Extruded Cut** parameters as shown in Figure 12.60 and accept these selections. Your part should look similar to Figure 12.61. *(Pay attention to the side that will be cut. You may need to flip the side to cut as the direction of the cut should be away from the majority of the spring.)*

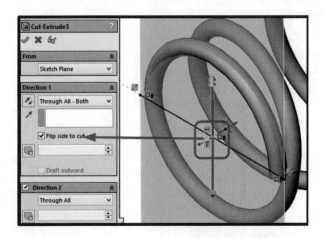

Figure 12.60

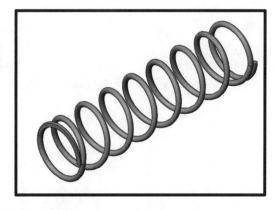

Figure 12.61

To see this being done, watch the video "**Cutting with a Line**."

- Use the **Back Plane** and the **Top** plane to cut the opposite end of the spring. This is how the spring will be after both ends have been ground to produce flat surfaces as shown in Figure 12.62 (***Do not** mirror the feature. Instead, try using a derived sketch to create this line.*)

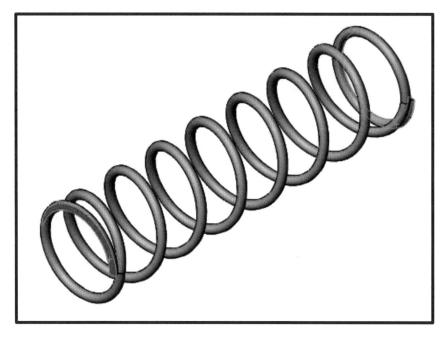

Figure 12.62

Axis Creation

- Hold the **CTRL** key and select the **Front** and the **Right** planes in the feature tree.

- Select **Reference Geometry, Axis**

- Accept the use of **Two Planes** to create the **Axis**
- Rename the default configuration to **Open**

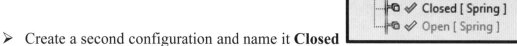

- Create a second configuration and name it **Closed**

Adding an Equation

> In the **Feature Manager Design Tree,** expand the **Sweep** and the **Composite Curve**:

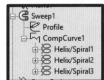

> Double click on the **Helix/Spiral 2** to show the dimensions in the graphics area.
> Name the **1.316** dimension **Distance**
> Again double click on the **Helix/Spiral 2** to show the dimensions in the graphics area.
> Double click on the pitch dimension (**0.188**) and delete the value field so that the **0.188** dimension is highlighted in blue 0.188in
> Type = (the equals sign)
> Double click on the **1.316** dimension in the graphics area to add it to the equation:

="Distance@Helix/Spiral2"

> At the end of this, type: /7
> Accept this equation-driven dimension:

To see this being done, watch the video "**Adding a Simple Equation**."

> In the feature tree, double click on your **Back End** plane.
> Right click on the **1.375** dimension and select **Configure Dimension**:

Configuration Name	Back End
	D1
Closed	0.280in
Open	1.375in

> Make the **Closed** distance **0.280** and accept this change
> Double click on the **Helix/Spiral 2** to show the dimensions in the graphics area.
> Right click on the **1.316** dimension in the graphics area and select **Configure Dimension**.
> Set the **Closed** distance to **0.221** and accept this change:

Configuration Name	Helix/Spiral2
	Distance
Closed	0.221in
Open	1.316in

➤ Check that each configuration works without any issues. Your two configurations should look similar to Figure 12.48 (as shown on page 258):

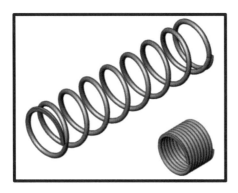

➤ Set the material to **1023 Carbon Steel Sheet**

Adding a Favorite Material

➤ In the feature tree, right click on your material and select **Manage Favorites**:

➤ Expand the **SolidWorks Materials**
➤ Expand the **Steel** section.
➤ Select the **1023 Carbon Steel Sheet (SS)**
➤ Select **Add**
➤ Close the material window

Configuring Material

➤ Right click on the material and select **Configure Material**
➤ In the **Modify Configurations** window, select the drop down menu in the **Material** column and select **1023 Carbon Steel Sheet** as the material for the closed configuration:

Configuration Name	Material
Closed	1023 Carbon Steel Sheet (SS)
Open	1023 Carbon Steel Sheet (SS)

➤ Accept this selection.

➤ **Save your work.**

Base

You will now create the Base, as shown in Figure 12.63. Later, this part will be altered in the context of the assembly.

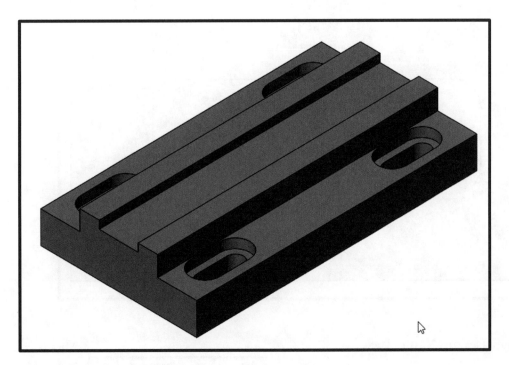

Figure 12.63

> In the front plane, create a mid-plane extrusion **4.5 inches** long, using the values shown in Figure 12.64.

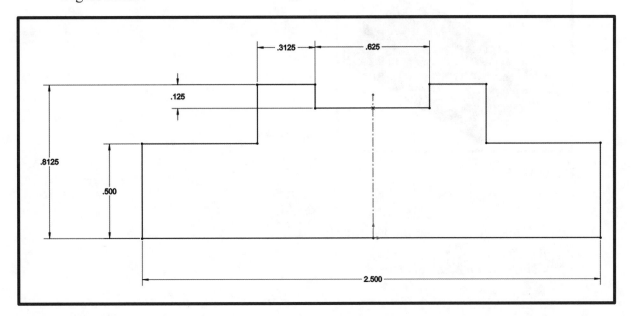

Figure 12.64

➢ Select the bottom face of your part and create an **Extruded Cut** 0.1 deep as shown in Figure 12.65.

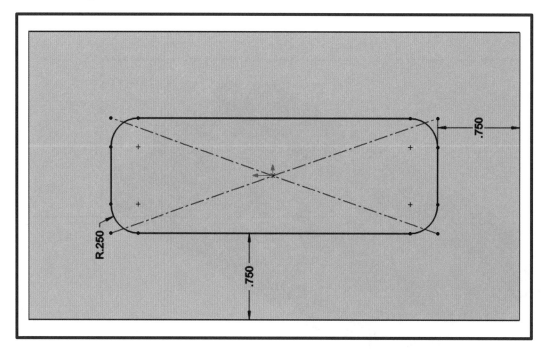

Figure 12.65

➢ **Save** this part as **YourName_Base.**

➢ Set the material to **AISI 1020 Steel, Cold Rolled**
➢ Select the face of your part as shown in Figure 12.66.

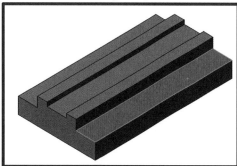

Figure 12.66

➢ Select **Hole Wizard** from the **Feature** tab
➢ Look **Normal To** this face.

Chapter 12: Centering Jig—Parts **269**

> Select **Counterbore Slot** as the hole type.

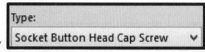

> Set the **Type** as **Socket Button Head Cap Screw**
> Set the remaining hole conditions as shown in Figure 12.67.

Figure 12.67

> Select the **Positon** tab.
> Create four points on the part as shown in Figure 12.68. *(Press the **ESC** key after the four points are created.)*
> Select the center line and add the **Vertical** constraint.
> Fully define the sketch as shown in Figure 12.69 and accept these settings

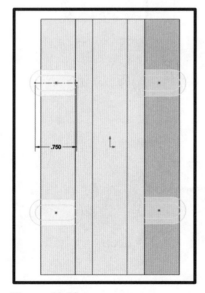

Figure 12.68

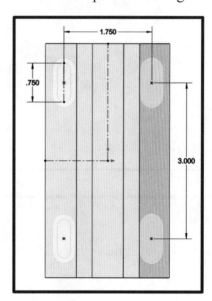

Figure 12.69

> **Save your work.**

Center Block

The **Center Block** as shown in Figure 12.70 allows the jaws to move symmetrically around the center of the jig.

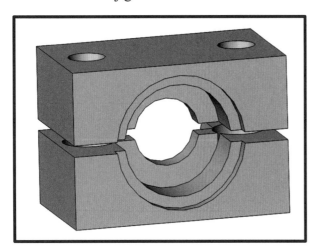

Figure 12.70

> In the front plane, create a center point rectangle as shown in Figure 12.71.

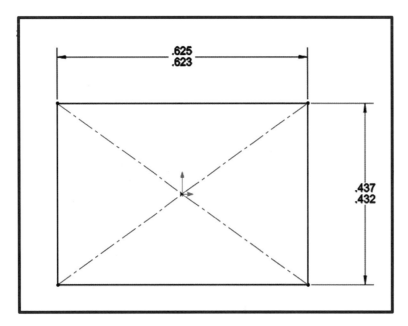

Figure 12.71

> Extrude this rectangle with a mid-plane extrusion that is 0.375 deep.

Chapter 12: Centering Jig—Parts **271**

> In the top plane, create a **Revolved Cut** feature. Make your sketch as shown in Figure 12.72 and set the **Revolved Cut** parameters as shown in Figure 12.73, using the line through the origin as the **Axis of Revolution**.

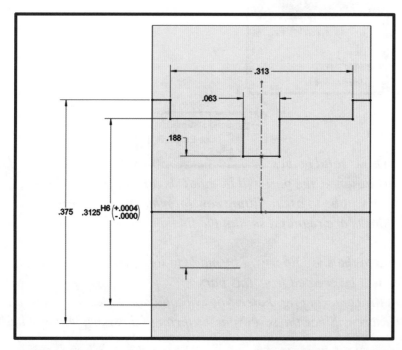

The H7 fit on the hole will allow for a light press fit between the bushing and the hole.

Figure 12.72

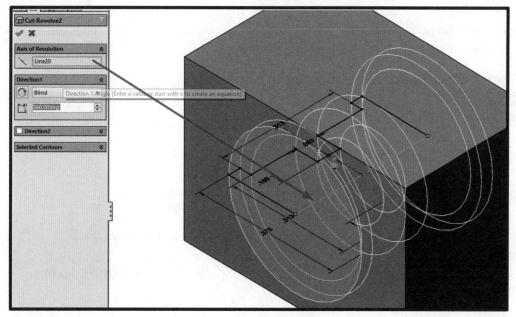

Figure 12.73

> Set the material to **Alloy Steel**
> **Save** your part as **YourName_CenterBlock**.

Two Solid Bodies in One Part

➢ On the front of the part, create an **Extruded Cut**, **1/32 inches** wide and symmetrical about the center, all the way through the part.
➢ When you exit the **Extruded Cut,** the **Bodies to Keep** window will open, choose to keep

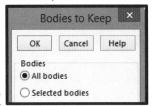

All bodies and select **OK**

Your feature tree will now show two solid bodies. *Most of the time, this should be avoided; however, in this instance, the part will be made as one block, and the last operation the machinist will perform will be to use a slitting saw (or something similar) to cut the block into two parts. The two blocks are required so that the final part can be properly assembled.*

The part drawing to be released to the shop floor will be made to look similar to this part. But the parts used in the assembly will be created from this part.

This part will hold the shaft in the centering jig. Later, you will add bolt holes to it so that it can be lightly clamped around the screw. Since the screw passes through this part, you will need to set up appropriate tolerances.

Bilateral Tolerance

This center block will be one of the main parts that gives the jig its accuracy to position all parts on center. Therefore, it will require a few tight tolerances.

➢ Edit the **0.063** dimension used in the revolve cut, by changing the **Primary Value** to **0.0625**.

➢ Select the **Bilateral Tolerance Type**

➢ Set the **Minimum Variation** to **-0.001 inches**

➢ Set the **Units Precision** to four decimal places
➢ Accept these selections.
➢ In your screw, edit the **0.063** dimension to a bilateral dimension of **0.0625 +0.001**:

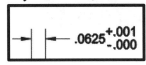

- In your screw, edit the **Ø0.188** dimension to **Ø0.188 -0.005**

- In the center block, edit the **0.188** dimension to **0.188 +0.005**

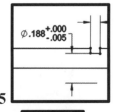

- Edit the **0.625** slot width in the base to **0.624 +0.002**

- **Save your parts**.

All mating holes will be added later in the assembly to ensure there are no alignment issues.

Derived Part

- Expand the **Solid Bodies** folder in the feature tree
- Right click on the lower solid body and select **Insert into New Part**
- Accept the settings in the **Insert into New Part Property.**
- If prompted "**Do you want to change the unit of measure of the derived part?**" answer **Yes**.
- Name this part **Upper Center Block**
- Select **Save.**
- Repeat this procedure for the **Lower Center Block**.

Your new part will open, and the feature tree will contain only one feature, a derived part feature . The arrow at the end of the derived feature name lets you know that this part was created in the context of another part.

To see this being done, watch the video "**Insert into New Part**."

Left Jaw

The right jaw will be a derived part from the left jaw as shown in Figure 12.74. The internal thread will be cut into this part using the thread on the screw, this will allow for an exact mating of the two parts threads.

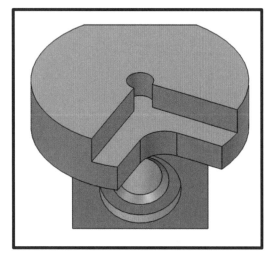

Figure 12.74

➤ Create the extrusion as shown in Figure 12.75

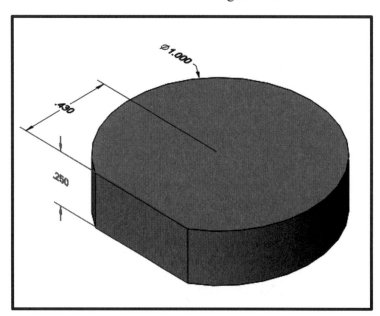

Figure 12.75

Degree Minute Seconds

- Select the **Options** from the top of the screen or from the **Tools** drop down menu.
- Select the **Document Properties** tab
- On the left side of the window, select **Units**
- In the chart, in the cell that lines up with the **Angle** row and the **Unit** column, select **deg/min/sec** as the angular units.
- Select **OK** to accept this change.
- Create a **1/8** deep **Extruded Cut** from the top of your part by creating the sketch as shown in Figure 12.76.

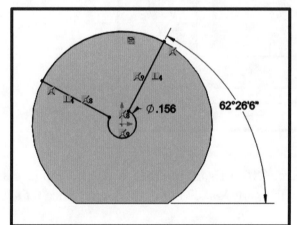

Remember, a sketch such as this must be a closed shape. Do not forget to create a co-radial, three-point arc or to convert the outer edge and trim the sketch.

Save your part as YourName_Jaws.
Set the material to Alloy Steel.

Figure 12.76

- Create a sketch as shown in Figure 12.77. Use this sketch for an **Extruded Cut** feature through the part.

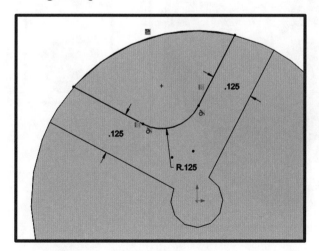

Figure 12.77

➤ On the bottom of your part, start an **Extruded Boss/Base** by creating the sketch as shown in Figure 12.78. Once your sketch is fully defined, exit the sketch.

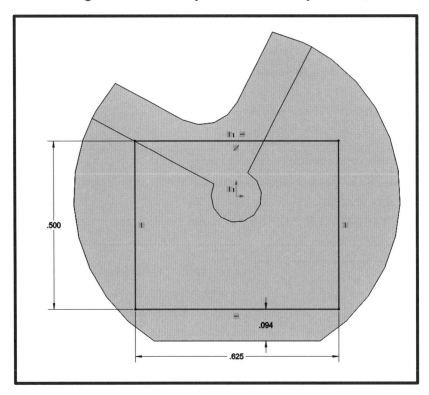

Figure 12.78

➤ For the extrusion **End Condition**, select **Offset From Surface**, select the top face as the surface, and set the direction to below the part. Make your extrusion distance **0.6875** as shown in Figure 12.79.

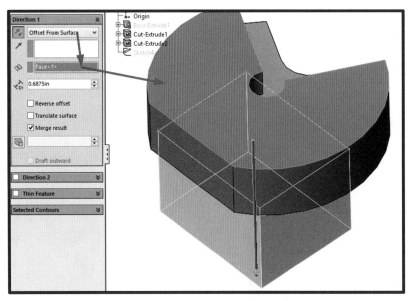

Figure 12.79

Add a 1/32 feature fillet to the four edges on the bottom circular face:

➢ In the right plane, create the revolved cut as shown in Figure 12.80.

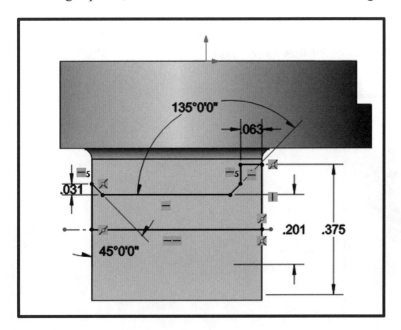

Figure 12.80

➢ Your final part should look similar to Figure 12.81.

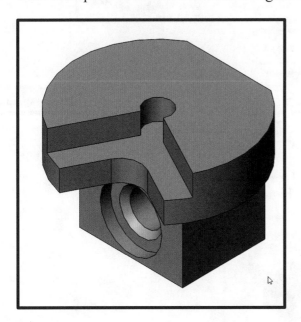

Figure 12.81

278   Chapter 12: Centering Jig—Parts

Right Jaw—Derived Part

> Create a plane offset 0.5 from the front plane as shown in Figure 12.82.
> Create an axis at the intersection of this plane and the right plane as shown in Figure 12.82.

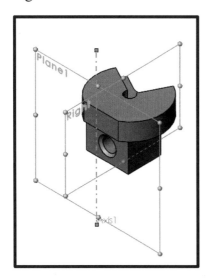

Figure 12.82

> Create a circular pattern, using this axis as the **Pattern Axis**

> Set the **Angle** to **180 degrees**

> Set the number of instances to **2** 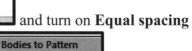 and turn on **Equal spacing**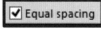

> Select the part as the **Body to Pattern**
> Your final part should look similar to Figure 12.83.

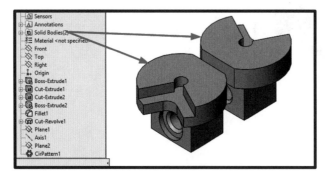

Figure 12.83

To see this being done, watch the video "Creating a Derived Part."

> Save your work.

> Using the **Insert into New Part** option, save the original jaw as **YourName_LeftJaw**. Save the derived jaw as **YourName_RightJaw**.

Edit in Context

Each of these jaws will require a sliding fit in the slot of the base. Adding a chamfer to the rectangular extrusion will allow for an easier assembly.

The dimensions required for the drawing are all found in the original part and not in the derived parts. This means the chamfer should be added to the original part so that its dimension will populate the drawing.

If you add a feature to the derived part, it will not be present in the parent part. But you need to start from the derived (the child part) to properly change the parent part and have it update the derived part.

> Open your right or left jaw solid part.
> Right click on the part in the graphics area or in the feature tree.
> Select **Edit In Context**

> Move the **Rollback Bar** to before the **Circular Pattern**
> Add a **0.05** chamfer to the part edges as shown in Figure 12.84.

> Edit the **0.625** dimension to a **Limit** dimension of **0.624 – 0.623**
> Move the **Rollback Bar** back below the **Circular Pattern**.

> **Rebuild** and **save** all three parts.

To see this being done, watch the video "Editing a Derived Part."

Name Plate

Often you will be required to include your customer's logo in a part. The wrap feature in SolidWorks makes this an easy task.

➤ Find a logo that you like and save it. You will be saving this part as **YourName_Plate**.

Figure 12.84

➤ In your top plane, create a rectangular extrusion that is **0.02 x 2 x 4 inches**.

➤ Set the material to **1060 Alloy.** This is an aluminium alloy

➤ Select the top face of your part and look **Normal To** it.

➤ Create a sketch on this face

➤ From the **Tools** drop down menu, select **Sketch Tools**, **Sketch Picture**

➤ Browse to where you saved your logo and open it.

➤ **Zoom to Fit**

➤ Select the grips on the corner of your image and resize it to fit inside the rectangle. *(You may also need to move it by clicking and holding the mouse button in and moving your mouse around.)*

➢ Resize and position your image until it is in the desired location, then exit the **Insert Sketch Picture**.

To see this being done, watch the video "<u>Inserting a Sketch Picture</u>."

To create usable geometry, you will need to sketch over the logo. Create fully defined geometry for your logo and exit the sketch.

Wrap

A wrap feature can scribe a sketch into a surface. It can emboss or raise the geometry of a sketch onto a part, or it can deboss or cut the geometry of a sketch into a part.

➢ Select the sketch from the feature tree.
➢ Select **Wrap** from the **Features** tab.
➢ Select the **Scribe** option in the **Wrap Parameters**
➢ Select the top face as the **Face for Wrap Sketch**
➢ Accept these selections.

To see this being done, watch the video "<u>Wrap Feature</u>."

Scale

➢ Create two more configurations of this part, called **Half** and **Three Quarter**:

➢ Make the **Three Quarter** configuration active.
➢ From the **Insert** drop down menu, select **Features** **Scale**
➢ Set the **Scale** to **0.25** **Uniform scaling**.
➢ **Scale about** the **Centroid**
➢ Accept these selections.

- Select the **Default** part configuration. If the **Scale** is not suppressed 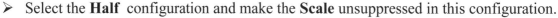, then suppress it in this configuration.
- Select the **Half** configuration and make the **Scale** unsuppressed in this configuration.

- Edit the **Scale** feature
- Set the **Scale Factor** to **0.5.**
- Select the **This configuration** radio button in the **Configurations** section:

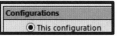

- Accept these selections.
- Verify that all three configurations update correctly.
- **Save your work.**

13: Centering Jig—Assembly

Screw Subassembly

- Start your assembly with your **Bottom Center Block**.
- **Save** this as **YourName_Screw Sub**.
- Add your screw and mate it in the orientation as shown in Figure 13.1. It should still rotate.
- Add the **Upper Center Block** and fully define it with mates in the orientation as shown in Figure 13.2.

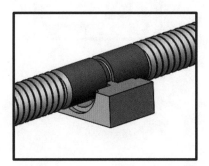

Figure 13.1

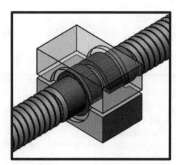

Figure 13.2

- Add a **Bushing**, fully defined with mates *(lock the rotation in the Concentric mate)*, to each side of the center block.
- Add a **Spring**, in the **Open** configuration, to each side of the center block as shown in Figure 13.3. *(Use a Concentric mate with the axis of the spring and lock its rotation.)*
- Add the **Left** and the **Right** jaws with the following three mates:
  - A **Concentric** mate between the hole in the jaw and the screw shaft.
  - A **Parallel** mate between the top of the jaws and the top of the center block.
  - A **Coincident** mate between the spring face and the Ø3/8 counterbored face as shown in Figure 13.4.

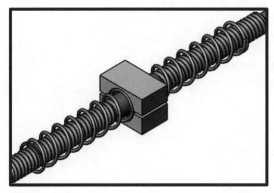

Figure 13.3

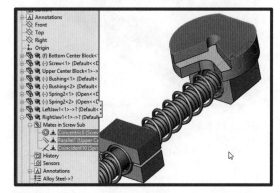

Figure 13.4

Assembly Configurations

- Select the **Configuration Manager** tab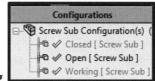
- Rename the **Default** 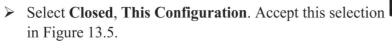 configuration to **Open**
- Add two more configurations named **Closed** and **Working**
- Select your **Closed** configuration
- In the **Feature Manager Design Tree,** select one of the springs.
- Select **Closed**, **This Configuration**. Accept this selection as shown in Figure 13.5.

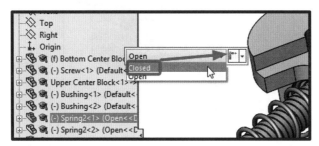

Figure 13.5

- Repeat this for the other spring. Your assembly should look similar to Figure 13.6.

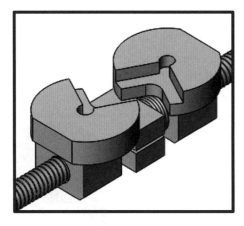

Figure 13.6

- Verify that both configurations update without any errors.
- **Save your work.**

Chapter 13: Centering Jig—Assembly

> Select your **Working** configuration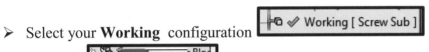

> Suppress 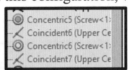 both springs in this configuration, which will also suppress the mates associated with the springs

> Select your **Open** configuration.
> Measure the distance between the two faces as shown in Figure 13.7.

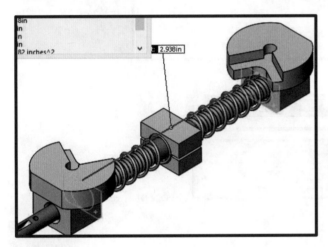

Figure 13.7

> Exit the **Measure** function.
> Select the closed configuration and repeat the measurement as shown in Figure 13.8.

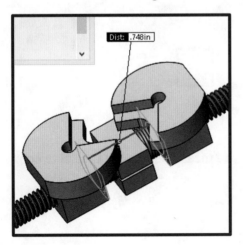

Figure 13.8

Chapter 13: Centering Jig—Assembly

Limit Distance Mate

- Select your working configuration.
- Add a distance mate between the two faces that you just measured.
- Make the distance **2.9375** as shown in Figure 13.9. **Do not accept this change yet.**

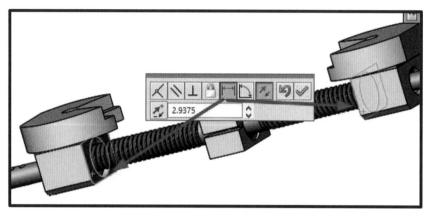

After setting up the limit distance mate, move the one jaw around and observe its behavior.

The limit distance mate should be working, but the assembly will not yet be behaving correctly.

Figure 13.9

- Select **Advanced Mates**

- Make the **Closed** distance **0.75** 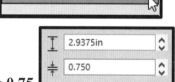. Accept these selections.

To see this being done, watch the video "**Creating a Limit Distance Mate**."

Symmetric Mate

- From the **Assembly** tab select **Mate**
- Select **Advanced Mates**.
- Select **Symmetric**
- As the **Entities to Mate,** select the two faces you measured the distance between:

➤ Select the **Front** plane of your **Bottom Center Block** as the **Symmetry plane** as shown in Figure 13.10. Accept these selections.

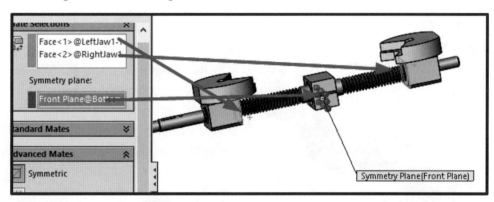

After this mate is complete, move one of your jaws and both should stay the same distance from the symmetry plane.

Figure 13.10

To see this being done, watch the video "**Symmetry Mate**."

Screw Mate

➤ From the **Assembly** tab, select **Mate**

➤ Select **Mechanical Mates**

➤ Select **Screw** as the mate type, select **Revolutions/in,** and set the value to **20**:

➤ Turn on the visibility of your **Temporary Axes**

➤ Select the axis of the screw and the circular edge of your left jaw as the **Entities to Mate** as shown in Figure 13.11 on the next page. Accept these selections.

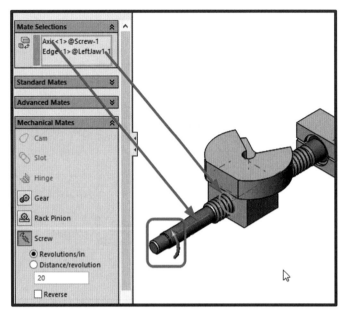

After finishing this mate, move your jaws again.

When the jaws move closer to one another, the screw should turn clockwise.

When the jaws move away from one another, the screw should turn counterclockwise.

If the screw rotates in the wrong direction, then edit the mate and check the box to reverse it, or right click on the Screw mate and select Reverse:

Figure 13.11

To see this being done, watch the video "Screw Mate."

➤ Verify that all three configurations still rebuild with no errors.

➤ **Save your work**.

Cavity

➤ Select your closed configuration.

➤ Click on your left jaw and select **Edit Part** . Do not open the part. Merely edit it in the context of the assembly.

➤ From the drop down menus, select **Insert**, **Molds** and **Cavity:**

➤ For the **Design Component Parts,** select the **Screw** from the graphics area or the feature tree

➤ **Do not** alter the **Scale Parameters** from **0%**.

➤ Accept these selections and **Return** to the assembly

To see this being done, watch the video "Using Cavity to Create Internal Threads."

- Select the **Right** plane of your **Bottom Center Block**.
- Select **Section View**
- Accept the selection. Your assembly should look similar to Figure 13.12.

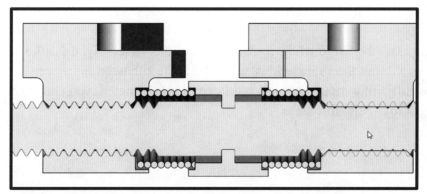

You should see a perfectly mated thread now in your left jaw and interference in your right jaw.

The cavity command will remove any material where the two parts interfere.

Figure 13.12

- Repeat these steps for the right jaw.
- **Save your work and** close the assembly.

Main Assembly

- Create an assembly with your **Base** as the first part.
- **Save** this as **Your Name_Centering Jig**.
- Insert the **ScrewSub, Working** configuration, and assembly, and fully mate the center block in the center of the base slot as shown in Figure 13.13 *(Try to add all mates to planes so that the part will stay centered if the design is changed.)*

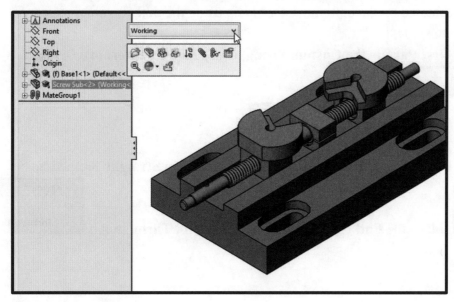

*Try to move the subassembly, it should not move. This is because SolidWorks will make it **Rigid** when it is first inserted so that it is easier to mate.*

Figure 13.13

Flexible Subassembly

> Click on the **Screw Sub** in the feature tree and select **Make Subassembly Flexible**
> 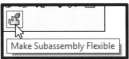.

SolidWorks will make an assembly rigid when you add it to another assembly, this allows you to add the required mates without the subassembly moving. If the subassembly were allowed to move it would be difficult to mate. Once it has been properly mated, you can allow it to move as design by setting it to the flexible state.

Hole Series

> Rotate your model around and select the undercut bottom face of the base:
>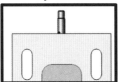

> Look **Normal To** this face.

> With the face still selected, add a **Hole Series** 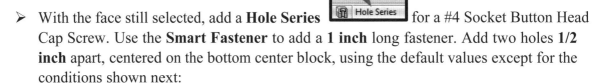 for a #4 Socket Button Head Cap Screw. Use the **Smart Fastener** to add a **1 inch** long fastener. Add two holes **1/2 inch** apart, centered on the bottom center block, using the default values except for the conditions shown next:
>   o For the **First Part,** in the **Custom Sizing,** change the **Counterbore Depth** to 0.015
>   o For the **Middle Part,** use a **Hole**
>   o For the **Last Part,** choose the **Upper Center Block**, a **Straight Tapped** hole 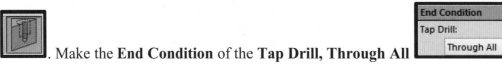. Make the **End Condition** of the **Tap Drill, Through All**

Your part should look similar to Figure 13.14.

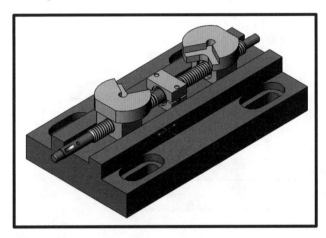

Figure 13.14

➢ **Save your work.**

*Sometimes when you use a hole wizard or a hole series and exit the command, all of the annotations will appear on the screen as shown in Figure 13.15. To clear these, right click on the **Annotations** folder in the feature tree and uncheck **Display Annotations**:*

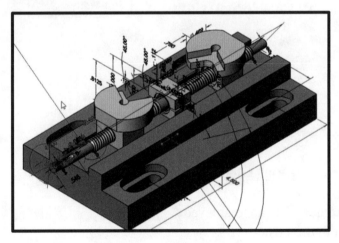

Figure 13.15

You should notice that the clearance holes, as shown in Figures 13.14 and 13.15 above, break out of the bottom center block. In this case, this would be acceptable, since it does not impede the function of the jig; however, it is not desirable. There is plenty of room to widen these slots, but when we do widen the slots, it would be nice if all three parts in the slots also update. This can be easily accomplished by following the next steps.

292 Chapter 13: Centering Jig—Assembly

Global Variable Creation

- Click on your **Base** and select **Open Part**
- From the **Tools** drop down menu, select **Equations**
- Set up the **Equations, Global Variables, and Dimensions** window as shown in Figure 13.16. **Do not exit this window yet.**

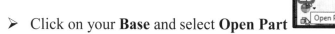

Figure 13.16

- Check in the box beside the **Link to external file**, found in the lower left-hand corner.
- In the **Link Equations** window, select the **Create new file** radio button:
- If the file location does not point to where you are saving your parts, then navigate to that folder. A file called **equations.txt** will be created in this location. In this file, it will be the global variable **SlotWidth** that you just created.
- Select **Link**
- Select **OK**

To see this being done, watch the video "**Creating Global Variables**."

- Double click on your part so the dimensions are displayed.
- Double click on the 0.624 slot width dimension.
- Delete the dimension and enter: = (the equals sign).
- Select **Global Variables, SlotWidth**:

- Validate the equation by clicking the green check mark at the end of it:

- Accept this selection.

Chapter 13: Centering Jig—Assembly **293**

➢ Rebuild your part. It should look similar to Figure 13.17

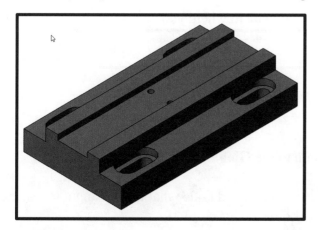

Figure 13.17

➢ Change your **1.75 inch** dimension for the center to center **C'Bore** distance to **2.0 inches** as shown in Figure 13.18. Also change the width of the base from **2.5** to **2.625 inches**.

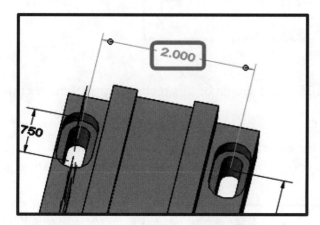

Figure 13.18

➢ Rebuild, save, and close the base.
➢ Open your left or right jaw solid part.
➢ Right click on the part and select **Edit in Context**
➢ From the **Tools** drop down menu, select **Equations**
➢ Check in the box beside the **Link to external file** , found in the lower left-hand corner.
➢ In the **Link Equations** window, select the **Link to existing file** radio button:

Chapter 13: Centering Jig—Assembly

- Click the **Browse** 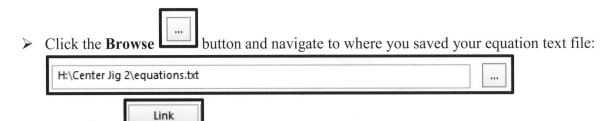 button and navigate to where you saved your equation text file:

- Select **Link**
- Select **OK**
- Now, link the **0.624-0.623** dimension to the **Global Variable, SlotWidth**.
- Delete the **Fillet** feature.
- Look at your parts from the top view. You should notice that the rectangular base extends slightly beyond the circular boss as shown in Figure 13.19.

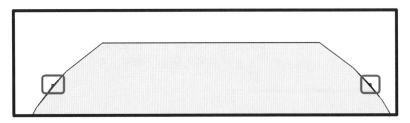

Figure 13.19

- Change the **0.094** dimension to **0.125** in the sketch of this rectangular feature to correct this problem.

- Move the **Rollback Bar** to before the circular pattern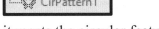
- Add a 1/64 fillet all around the rectangular base where it meets the circular feature, as shown in Figure 13.20.

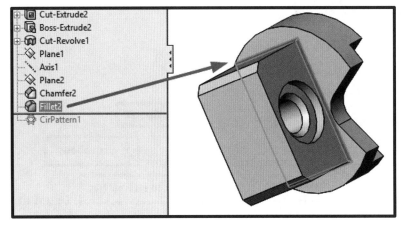

The fillet at this location is important as it will minimize the chance of this part cracking during heat treatment.

Figure 13.20

- Rebuild, save, and close all jaw parts.

Chapter 13: Centering Jig—Assembly **295**

➢ Rebuild your main assembly. You should see a warning message similar to that shown in Figure 13.21. Close this message.

Type	Feature	Preview	Help	Description
⚠ Warning	📋 Smart Fastener1			Hole series has changed. Please verify that current fastener is correct.

Figure 13.21

➢ Right click on the **Smart Fastener** folder at the bottom of the feature tree and select **Mark up to date** [Mark up to date]. This will dismiss the warning.
➢ Rebuild and save this assembly.
➢ Repeat this procedure for the center block parts. You will need to edit both halves in context, or the assembly will have errors.

The center of the center block is where the center of all parts should be if this jig is designed and built correctly. Currently, it is held on the center of the slot between the sides but not between the ends.

If this jig were to require repair and the center block was unbolted, its position when reattached could vary from end to end because of the clearance holes. Therefore, we should redesign this part so that if it is disassembled and reassembled, it will locate in the same position (+/- 0.0005 or less) each time.

There are several ways to do this. One is to add a dowel pin to lock up the location of this part. Another is to make this part have clearance in the slot and add two dowels. But as designed, the only way to disassemble the center block is to remove the jig from the machine table, meaning that the jig will need to be re-indicated once repaired. So a simple solution is to add a dowel hole in the center of the top center block, which can be easily indicated.

➢ Add a 1/8 dowel hole to the top of the center block as shown in Figure 13.22

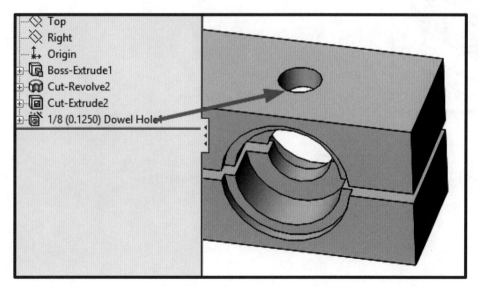

Figure 13.22

In-Context Parts

As the jig is currently designed, it would be awkward to use, because the subassembly could easily lift up on the ends while in use. To prevent this, we will add some O1 flat bar parts to hold the jaws in place and prevent them from riding up.

> From the **Assembly** tab, select the drop down arrow below **Insert Components**. Then select **New Part**
> Click in the graphics area, and you will have created a new blank part. If you look in the feature tree just above your mates folder, you will see the new blank part that is fixed in the assembly
> Click on this part in the feature tree, and select **Edit Part**
> Select the top face of your base as shown in Figure 13.23.

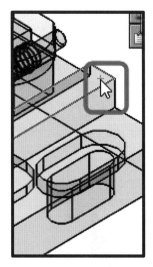

By selecting this face, you are establishing the plane on which this new part will be created, in the context of the assembly. This is also known as Top Down Design.

Figure 13.23

> From the **Features** tab select **Extruded Boss/Base**.

Chapter 13: Centering Jig—Assembly

➢ Create the sketch as shown in Figure 13.24 and extrude it up off the base by **1/8 inch**.

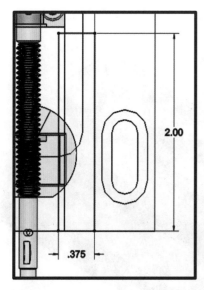

*When we create an in-context part such as this, the default setting allows the part to be constrained to other parts within the assembly as shown here. These are known as **External References**. They will fully define our part in the assembly, and this parts origin will be the assembly origin.*

*If you do not want this, you can **Float** the part after it is finished. Or if you know you do not want the external references before you start the sketch, then you can select **No External References*** *and then fully define your sketches as you normally would.*

Figure 13.24

To see this being done, watch the video "**Creating an In Context Part**."

➢ Exit the editing of this part so that you are back in the assembly and no parts are see-through.
➢ **Rename** this part **YourName_Hold Down**.
➢ Set the material to **O1**.
➢ Save the assembly. A new **Save As** window will open asking if you want to save the part internally or externally as shown in Figure 13.25. Save externally.

Figure 13.25

If you choose to save the part internally, you will not be able to easily create drawings of this part, but you can always save it externally latter. If you choose to save it externally, you will be asked to navigate to where you would like the part saved, and drawings can be easily created.

298 Chapter 13: Centering Jig—Assembly

➢ Create a linear component pattern of your **Hold Down** so that there are four parts as shown in Figure 13.26. You should be able to calculate the distances required.

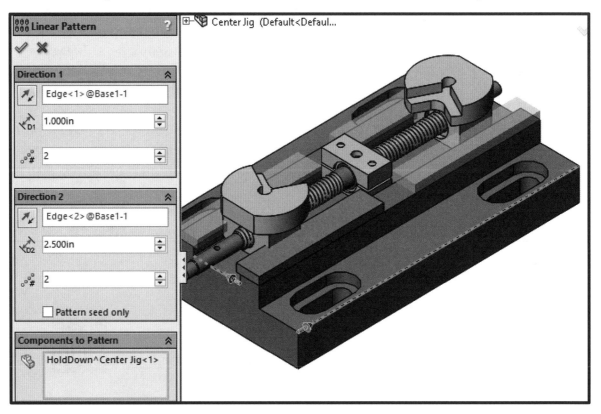

Figure 13.26

➢ Add a hole series through your original **Hold Down** and the **Base. Do not** add a smart fastener. Create the holes, on center of the **Hold Down**, *1/4 inches* from the one end and *1 1/2 inches* apart. Use the **Countersink** hole type 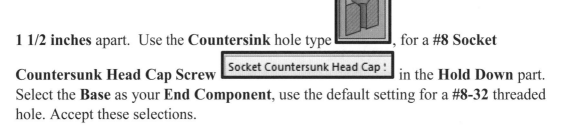, for a *#8 Socket Countersunk Head Cap Screw* in the **Hold Down** part. Select the **Base** as your **End Component**, use the default setting for a *#8-32* threaded hole. Accept these selections.

➢ **Save your work.**

In-Context Holes

➢ Edit the base, but **do not open it and do not use the hole wizard** from the assembly hole series.
➢ Select the top face that the **Hold Down** parts are on and look **Normal To** it.

➢ With the face selected, add **#8-32 Tapped, Hole Wizard** holes at the center of the three remaining **Hold Down** parts as shown in Figure 13.27.

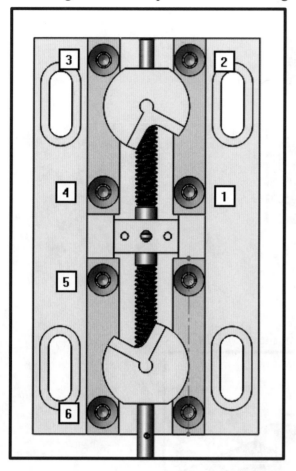

*When adding the holes, if you place your mouse near the edge of a hole, it will "wake up" the center of the clearance hole. You can then place the point on the center of this hole, adding the in-context **Coincident** mate to lock the tapped hole center in line with the clearance hole*

Now, if you change the clearance hole location, the tapped hole location will also change, and the holes will always be perfectly aligned.

Figure 13.27

To see this being done, watch the video "<u>**Adding in Context Hole Wizard Holes**</u>."

➢ Add **8** (fully mated), **#8-32 x 5/16 long Socket Countersunk Head Cap Screw**s

➢ Use the **Measure** tool  from the **Evaluate** tab to calculate the interference between the jaws and the hold downs.

➢ Edit the jaws in context to create a slot that has a clearance of **0.015 inches** all around the hold down as shown in Figure 13.28.

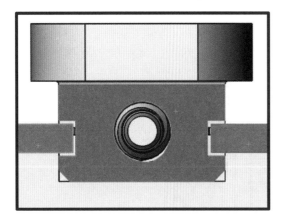

Figure 13.28

➢ **Save your work.**

➢ Select the end of the base as shown in Figure 13.29

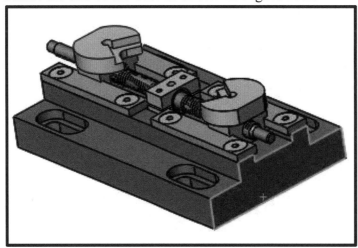

Figure 13.29

➢ Insert another new part. **Name** this **YourName_End Support**.
➢ Make the material a **Gray Cast Iron**.
➢ Create this part from the end you just selected as the sketch plane.

> Create the pink sketch as shown in Figure 13.30, using the convert entities and offset commands where possible. *(The sketch is shown in pink to make it easier to see the sketch entities, but your sketch will have only black lines once it is fully defined.)* The 1/64 inch dimension is from the surface of the jaw, and the 1/32 inch dimension is to the diameter on the screw.

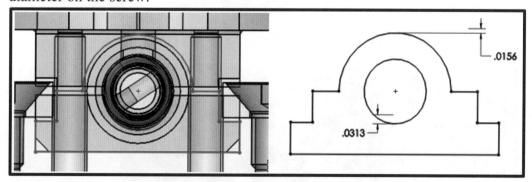

Figure 13.30

Figure 13.31 is angled to help you visualize the required profile.

Figure 13.31

> Extrude the sketch 5/16, and your assembly should look similar to Figure 13.32

Figure 13.32

> Insert another new part. **Name** this **YourName_Nylon Bushing.**

302 Chapter 13: Centering Jig—Assembly

- Make the material **Nylon 6/10.**
- The bushing will be 5/16 long, its OD (outside diameter) will equal the ID (inside diameter) of the end support, and its ID will equal the OD of the screw at this end.
- Your assembly should now look similar to Figure 13.33.

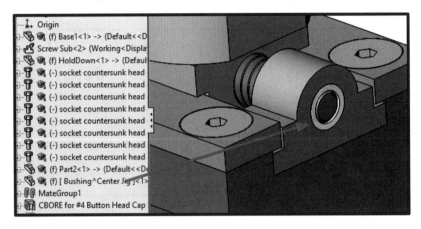

Figure 13.33

- **Save your work.**

To see this bushing being created, watch the video "**In Context Nylon Bushing**."

Adding a Folder

By now, you have a number of fasteners in the feature tree. They are adding to the clutter and slowing you down as you search for and edit parts. These can all be placed in a folder.

- Select all of the fasteners in the feature tree.
- Right click and select **Add to New Folder** as shown in Figure 13.34. Name this folder **Fasteners**

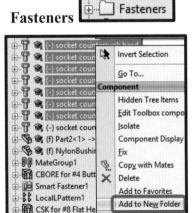

As you add more fasteners, they can be dragged and dropped into this folder.

Figure 13.34

To see this being done, watch the video "**Adding a Folder for Fasteners**."

Up to Vertex Extrusion

- Open your **End Support** part.
- In the middle of your part, create the sketch as shown in Figure 13.35.
- You may need to create a mid-plane. Extrude this using the up to vertex option, in both directions, as shown in Figure 13.36.

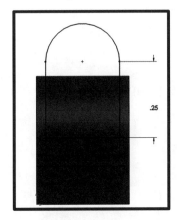

Figure 13.35

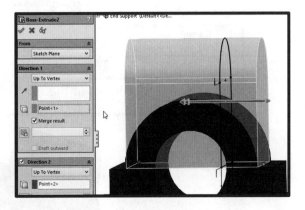

Figure 13.36

To see this being done, watch the video "**Up to Vertex Extrusion**."

- Add a 1/32 wide **Extruded Cut** as shown in Figure 13.37.

Figure 13.37

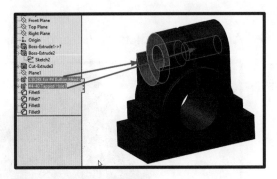

Figure 13.38

- Add a counterbored hole for a #4 button head cap screw and a #4-40 tapped hole as shown in Figure 13.38. Both holes are on center of the arch.

A clearance hole on one side of a slot and a tapped hole on the other side of the slot allows this part to clamp on the shaft, which passes through the hole. Do not set your end condition of your clearance hole to through all, as this note on the drawing will instruct the manufacturer to drill out the tapped half, rendering this feature useless. Instead, set the end condition to up to next. Do not worry about the edge left around the clearance hole, as the manufacturer will file it so that it is not sharp.

Face Fillet

- Select **Fillet** from the **Features** tab.
- Select **Face Fillet** 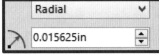 as the **Fillet Type**
- Set the **Fillet Parameters** to create a **Radial 1/64** fillet:

- Select the faces as shown in Figure 13.39. Accept these selections.

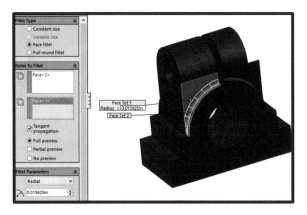

Figure 13.39

- Add four face fillets as shown in Figure 13.40.

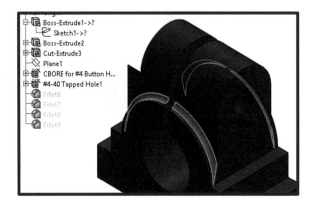

Figure 13.40

To see this being done, watch the video "Face Fillet."

- **Save** and close this part.
- Add an appropriate hole series and fastener to secure this part to the base.

Completing the Assembly

- Add a Ø1/16 x 3/8 long split pin mated to the hole and centered on the shaft:

  o From the **Design Library** **Toolbox**  select **Pins, All Pins:**

  o Drag and drop a **Spring Slotted Pin**  into your part.
  o Use a **Concentric** mate with the rotation locked and a **symmetry** mate between the ends of the pin and a plane in your screw.
- Add a **#8 Narrow Flat Washer, Type B** to the shaft using a **Concentric** and a **Tangent** mate:

  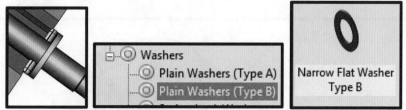

- Add a **#202 Woodruff Key** to the shaft using a **Concentric**, **Coincident**, and **Parallel** mate

  to the screw
- Add your **Handle** to the assembly and fully mate it in place using a **Concentric** and two

  Coincident mates

- Add another flat washer with a **Concentric** and **Coincident** mate

- Add a fully mated **8-32 Machine Screw Hex Nut**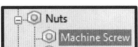

- Add a fully mated **4-40 x 5/16 Socket Button Head Cap Screw** to the end support:

- Add the handle grip, fully mated. Your assembly should similar to Figure 13.41.

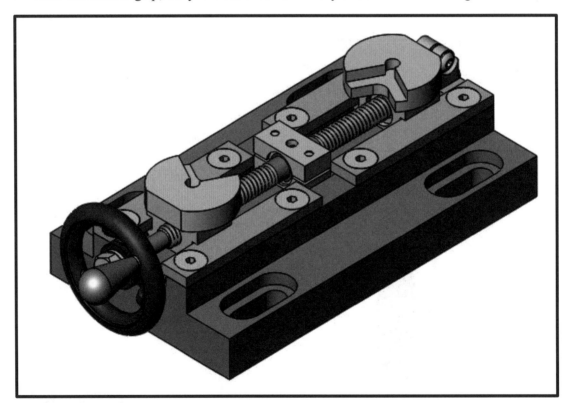

Figure 13.41

- Drag and drop all of the added washers, the woodruff key, screws, the nut, and the roll pin into the **Fastener** folder.

Assembly Configurations

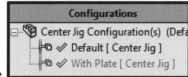

- Add a configuration to your assembly, named **With Plate**
- For the **With Plate** configuration, add the **Plate** in its **Half** configuration. Suppress this part in the default configuration.
- Mate the plate to the jaws to show the plate clamped as shown in Figure 13.42.

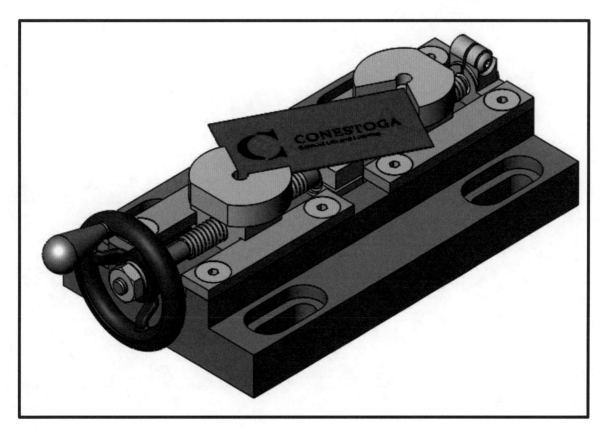

Figure 13.42

- **Save your work.**

With this style of jaw, there will be times when a part is not held properly. One way to correct this is to redesign the jaws as a two or more part assembly, which would allow the jaws to rotate and compensate for any alignment issues.

14: Turkey Call

This turkey call is slightly different than the one you will design. It was made as a prize for an archery competition to raise money for Breast Cancer research and has been used in the field to successfully call in wild turkeys.

Figure 14.1

Creating The Revolved Part

> Create a revolved part in the front construction plane using the drawing shown in Figure 14.2. Select **Maple** as the material type. Work in inches for this part.

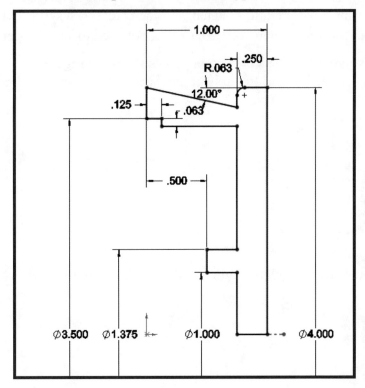

Figure 14.2

310 Chapter 14: Turkey Call

- Add a 1/16 fillet as shown in Figure 14.3.
- Add a 1/32 fillet as shown in Figure 14.4.

Figure 14.3

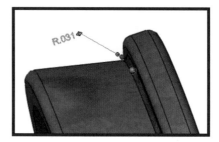

Figure 14.4

In the bottom of this part, we need to create an **Extruded Cut** in the shape of the pink ribbon that is used to show support for breast cancer research. The bottom of the part may not be your bottom view.

You will need to download the JPG file needed for this exercise to your computer's hard drive. This link will take you to the page to download the file <u>ribbon.jpg</u>.

Inserting the Sketch Picture and Creating the Spline

For this exercise, you will be inserting a picture into a sketch, and then creating a spline over the picture. This exercise is one of the few times when your sketch will not need to be fully defined.

- Click on the back face of your part and select **Normal To** as shown in Figure 14.5. This will orient the part so that you are looking squarely at the face that the sketch will be created on.

Figure 14.5

- Click on the face again and select **Sketch** to begin creating a sketch on this face as shown in Figure 14.6.

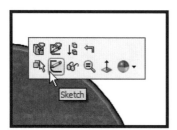

Figure 14.6

Chapter 14: Turkey Call **311**

➢ Select **Tools, Sketch Tools, Sketch Picture** as shown in Figure 14.7.

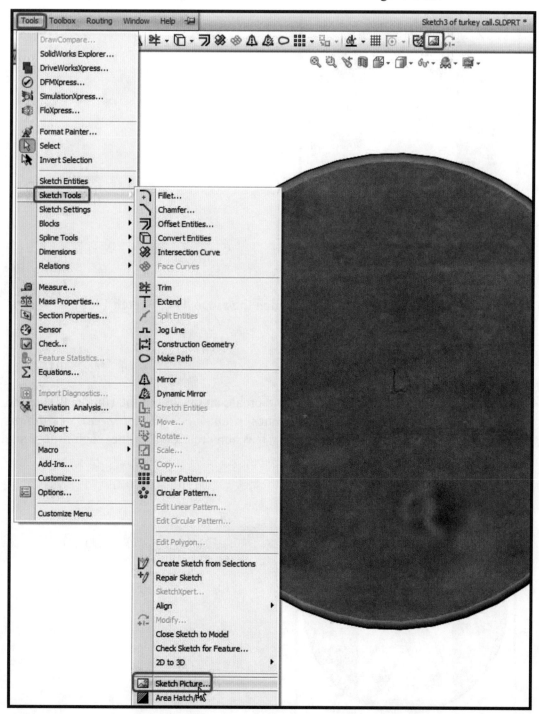

Figure 14.7

➢ Navigate to where you saved the file **ribbon.jpg** and select **open**. The image should come in very large, with the lower left-hand corner at the origin.

312 Chapter 14: Turkey Call

➢ Resize the image and center it (by eye) so that it looks similar to Figure 14.8.

Figure 14.8

To see the sketch picture being inserted, watch the video "<u>TC Sketch Picture</u>."

Creating the Spline

➢ Use the **Spline** sketching tool and create an outline of the ribbon as shown in Figure 14.9, creating three different splines. *(Remember, there is no exact solution or exact shape, this is simply for aesthetics.)* When creating a spline, each time you click, a control point will be added.

Figure 14.9

These control points can be used later to alter the shape by clicking on them and dragging them, until you are happy with the final shape. They can also be added or removed later by using the right mouse click options.

Chapter 14: Turkey Call   **313**

To see this being done, watch the video "TC Spline."

➤ Exit your sketch and use the sketch to create an **Extruded Cut**, **1/16 inch** deep into the part, as shown in Figure 14.10. Name this extrusion **Ribbon**.

Figure 14.10

At this point, you may not like the appearance of the ribbon. The next step will be to alter its appearance until you are happy with the look of it. (The whole purpose of this feature is to make the final part look nice and has nothing to do with function.)

*Since we will be working with this face a lot, it will make it quick and easier to select this face if we create a view orientation that is **Normal To** this surface.*

New View Orientation

➤ Click on this face and select **Normal To** as shown in Figure 14.11.

Figure 14.11

➤ Select **View, Orientation, Modify**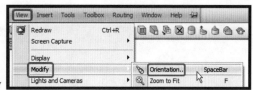

314 Chapter 14: Turkey Call

> Select the **New View** icon

> Call this new view **Ribbon** and select **OK**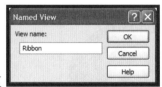

> **Save your work.**

To see this being done, watch the video "TC Create View."

*Now, whenever you want to work with or view the part from this new view, you can select it from the **View Orientation Menu** as shown in Figure 14.12.*

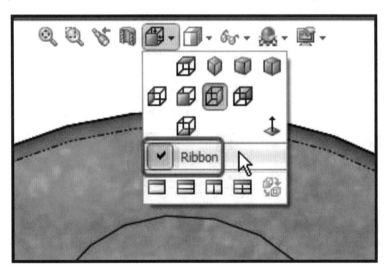

Figure 14.12

Editing the Spline

➤ Open the ribbon sketch up so that you can edit it.

➤ Click on a spline. You should see something similar to Figure 14.13.

*There are several ways to control the shape of a spline. A control point (#1) can be added or removed by right clicking on the spline; it can be selected and dragged around to alter the shape of the spline. The diamond handle (#2) can be used to rotate the spline about the control point. The arrow head (#3) can be used to drag the spline and change the tangent weighting point on that side of the control point. (Holding the **ALT** key while altering this will keep the deformation symmetrical about the control point.) Altering the spline using the circular handles (#4) will have the same effect as moving handles #2 and #3 at the same time.*

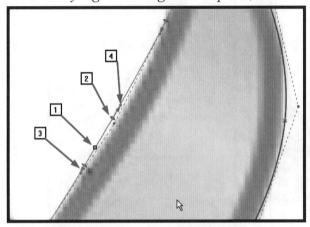

Figure 14.13

With any of the four control mechanisms, if you do not recall what their function is, just position your mouse over them. A graphic symbol of the control point's function will appear.

➤ Alter your splines until you are happy with the look of them. *Once you are satisfied with the appearance of your spline, it can be fully defined.*

➤ Add the vertical relationship to the four points shown in Figure 14.14 and to the origin.

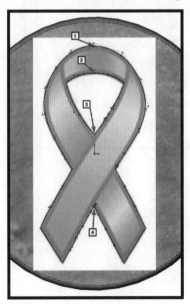

Figure 14.14

316 Chapter 14: Turkey Call

➢ Select **Tools, Dimensions, Fully Define Sketch.**
➢ Set up the **Fully Define Sketch** tab as shown in Figure 14.15. Then select **Calculate.**

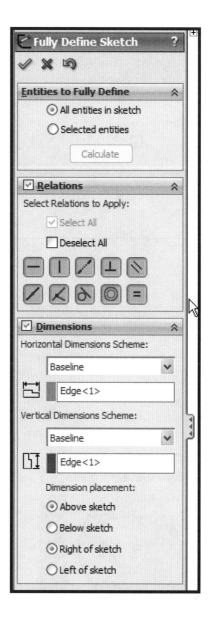

Figure 14.15

➢ This should fully define your sketch. You can exit this function and exit the sketch.

➢ **Save your work.**

Creating Work Planes

Currently, the ribbon feature looks reasonable. However it could look a lot better if it was given the appearance that one end of the ribbon is overlapping the other. This can be done several different ways.

A simple way to give the ribbon the desired appearance (so that it appears as if one end of the ribbon is overlapping the other end) is to create two new angled work planes for the sketch.

> Sketch on the surface as shown in Figure 14.16.

Figure 14.16

> Create a sketch, using only a center line that connects the two corners of the ribbon as shown in Figure 14.17.

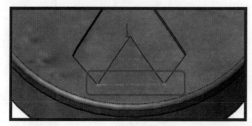

Figure 14.17

> Create a reference plane. Select the surface as shown in Figure 14.16 and the **center line** from the sketch shown in Figure 14.17 as the **Reference Entities**.

> Select to create this plane on an angle of **0.5 degrees** and accept these selections. The plane should cut into the part as shown in Figure 14.18.

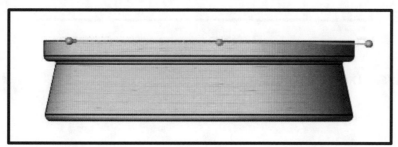

Figure 14.18

> Name this reference plane **First Angle Plane**.
> Create a sketch in the **First Angle Plane**.

Chapter 14: Turkey Call

> Create a center line that is **Horizontal** and **Tangent** to the top of the ribbon as shown in Figure 14.19.

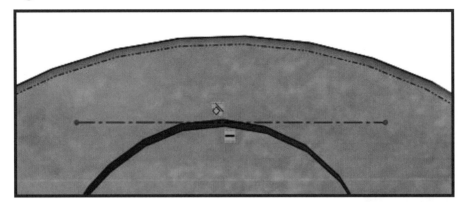

Figure 14.19

> Create a plane parallel to the top surface with this newly created center line sketch as the **Second Reference** as shown in Figure 14.20.

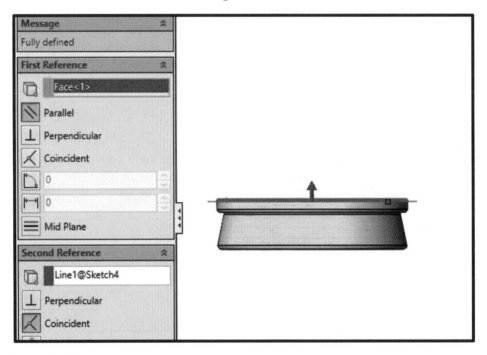

Figure 14.20

> Name this plane **Lower Parallel.**

Chapter 14: Turkey Call 319

➤ Create one more work plane called **Second Angle Plane.** Selecting the **Lower Parallel** plane and the last center line that you created, make this an angled plane, using **0.5 degrees** as the angle. Accept these selections, as shown in Figure 14.21.

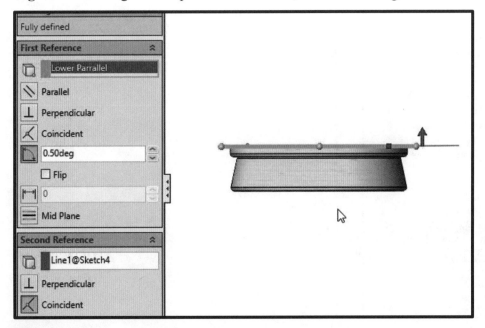

Figure 14.21

Repairing Sketch Errors

At this point, you are about to make some changes to your part that will cause problems to ripple through most of the planes that you have created.

➤ Use the **Save As Copy** option and save this part as **Turkey Call 2**.

➤ Right click on the **Ribbon** feature and select **Delete** ✕ Delete.... *(Do not delete the sketch, just the **Extruded Cut**.)*

➤ Set up the **Confirm Delete** as shown in Figure 14.22. And accept these selections. *(Do not delete the child features.)*

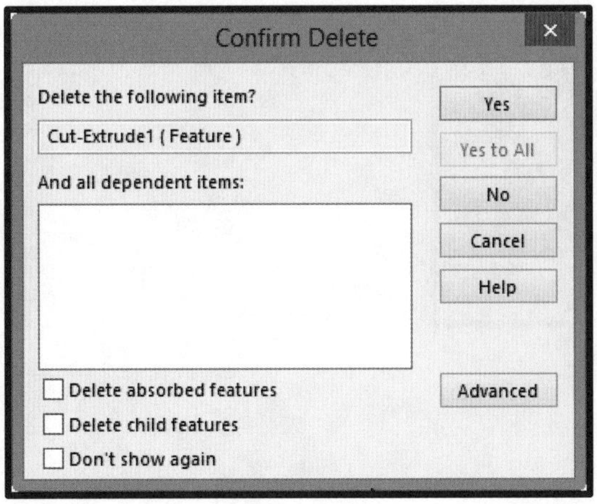

Figure 14.22

320   Chapter 14: Turkey Call

➢ From the error message window, select **Close**
➢ Problems will show up in the **Design Tree** as shown in Figure 14.23.

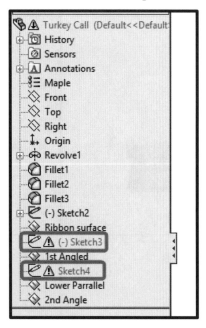

Figure 14.23

➢ Click on the first sketch with rebuild issues and select **Edit Sketch**
➢ You should see some relationships whose color has changed to **olive green**, meaning that the original geometry has changed, as shown in Figure 14.24.

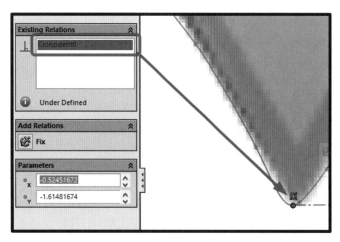

Figure 14.24

➤ Select any and all of these relationships and delete them.

To see this being done, watch the video "**Deleting Constraints**."

➤ Repair your sketch.
➤ Repair all of the remaining problems and **save your work**.

Creating the Layered Ribbon

➤ Create an **Extruded Cut** in the **First Angle Plane.**

➤ Click on the ribbon sketch and select **Show**

➤ Use the **Convert Entities** tool to convert your spline.
➤ Add center lines and trim the spline until it looks similar to Figure 14.25.
➤ Complete your sketch until it is closed in a manner similar to Figure 14.26.

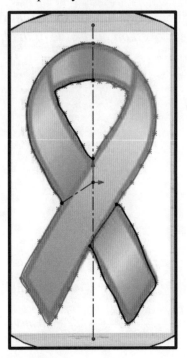

Figure 14.25

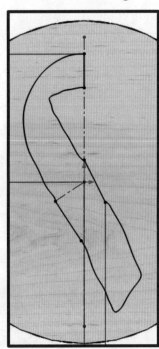

Figure 14.26

➤ Use the **Fully Define Sketch** tool again.

- The **Through All** end condition should be used for your extrusion. Your part should now look similar to Figure 14.27.

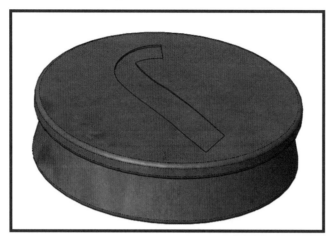

Figure 14.27

- Name this feature **Ribbon First Half**.
- **Save your work.**
- Repeat the above steps to create the other half of the ribbon, this time in the **Second Angle Plane**. *(Note: this half of the ribbon needs to be broken on each side of the first half and must not pass through it.)*
- Your part should look similar to Figure 14.28.

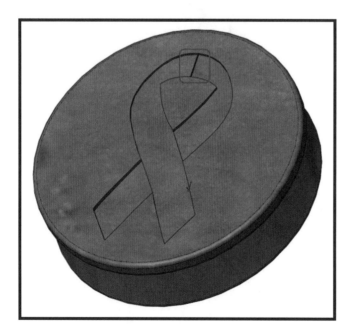

Figure 14.28

- **Save your work.**

Deleting a Face

There may be a thin wall of material, representing a face of zero thickness, where the two cuts meet. There are several ways to remove this feature. One is to go into the first **Extruded Cut** sketch and extend the area just past the center line. Another way to correct this feature is to delete the faces of each extrusion, since there is no real face here.

> Select **Insert, Face, Delete**

> Select only the face where the two cuts meet. The face of each cut will need to be selected.
> Choose the option to **Delete** and accept your selections.

To see this being done, watch the video "**TC Delete Face**."

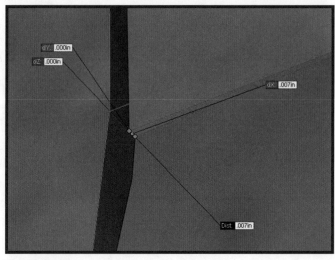

If you zoom in very close to where the two surfaces meet, you will notice that there is a very small step as shown in Figure 14.29. (Here it is 0.007, but your step may be a different value.) This is a very small step and can be ignored. Since the part will be cut into wood, this can easily be sanded out.

Figure 14.29

> Use the **Measure** tool under the **Evaluate** tab to calculate the distance of the step:

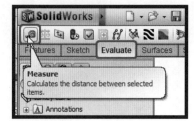

> **Save your work.**

Appendixes

Appendix 1: Constraints/Relationships

When adding relationships, the number beside the constraints in the graphics area indicates the entity number to which the relationship has been added. To turn the visibility on/off of the constraints select **View**, and then select **Sketch Relations**.

The following list does not include all possible relationships, just those used in this book.

Coincident—makes two or more entities on top of each other, such as the center of a circle may be coincident to the part origin.

Collinear—makes the endpoints of two or more lines lie in the same straight line.

Concentric—makes two or more arcs/circles share a center point.

Coradial—makes two or more arcs/circle have the same radius.

Equal—sets two or more entities equal, in length or in diameter.

Fix—locks an entity to its current location.

Horizontal—makes a line horizontal.

Intersection—locks a point at the intersection of two or more entities.

Merge—brings two or more points or endpoints together.

Midpoint—locks an endpoint to the center of another object.

Appendixes: Constraints/Relationships *(continued)*

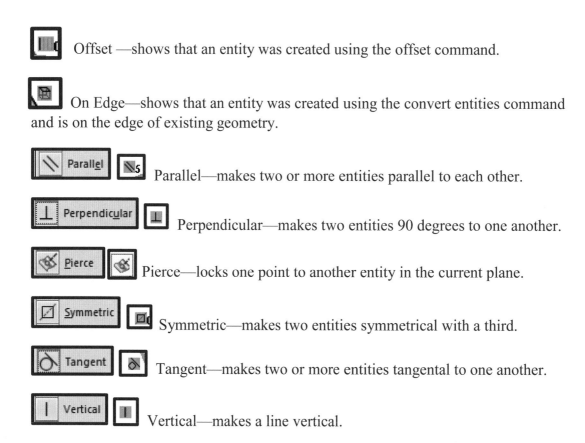

Offset —shows that an entity was created using the offset command.

On Edge—shows that an entity was created using the convert entities command and is on the edge of existing geometry.

Parallel—makes two or more entities parallel to each other.

Perpendicular—makes two entities 90 degrees to one another.

Pierce—locks one point to another entity in the current plane.

Symmetric—makes two entities symmetrical with a third.

Tangent—makes two or more entities tangental to one another.

Vertical—makes a line vertical.

Appendix 2: Dimension Tolerance Types

None — None—no dimension tolerance is given; the general tolerance table will control this feature size based on the number of decimal places.

Basic `15.90` Basic—the theoretical exact size of the dimension, the dimension will be enclosed in a rectangle.

Bilateral $15.90^{+0.01}_{-0.01}$ Bilateral—displays the plus and the minus tolerances for the feature.

Limit $\begin{array}{c}15.91\\15.89\end{array}$ Limit—displays the maximum and the minimum feature sizes.

Symmetric 15.90 ± 0.01 Symmetric—displays the basic feature size with the plus and minus being equal.

MIN `15.90 MIN` Min—add this to state that the dimension is the minimum feature size.

MAX `15.90 MAX` Max—add this to state that the dimension is the maximum feature size.

Fit $15.90\ g8$ Fit—displays the basic feature size and the tolerance grade.

Fit with tolerance $15.90\ g8\ ^{-0.01}_{-0.03}$ Fit with Tolerance—displays the basic feature size and the tolerance grade with the limits of the tolerance displayed.

Fit (tolerance only) $15.90^{-0.01}_{-0.03}$ Fit (tolerance only) —displays the basic feature size and the limits of the tolerance grade but not the number and letter of the tolerance grade.

Appendix 3: Short Cuts

When sketching geometry, there are a number of short cuts that can be accessed by selecting a sketch entity. These short cuts include:

 The vertical, horizontal, and fixed relationships.

 Select Other—using this when entities overlap will allow you to toggle through which entity you wish to select.

 Construction Geometry—will change an entity to or from construction geometry.

 Zoom to Selected—will fit the selected entity (entities) to the screen.

 Replace Selected—will let you replace the selected entity with another existing entity.

Sketch Numeric Input—this feature can be toggled on and off; it allows you to enter the

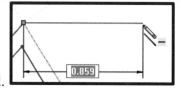

entity size as it is sketched.

 Smart Dimension—the standard dimensioning tool.

 Exit Sketch.

Appendix 4: Built-In Calculator

SolidWorks has a useful built-in calculator. Any field where a number can be entered also can accept simple equations. Mathematical functions can be called, and units can be specified in these fields. All you have to do is enter the equation and press the **Enter** key, and SolidWorks will perform the calculation.

The calculator accepts arithmetic operators, as well as the characters for plus (+), minus (–), divide (/), and multiply (*). The use of parentheses to create a more complex equation as well as standard algebraic notation are accepted. The equals sign (=) is not required when using this function.

Note: When entering the unit short form after a value, the value will be converted to the current unit settings. For example, if you are working in inches and you enter 25.4 mm, the field will change to 1.00 inches after the **Enter** key is pressed or a new field has the focus.

The unit short forms shown in Table 1.1 can be used.

Symbol	Description	Symbol	Description
deg	Degrees	m	Meters
'	Minutes	km	Kilometers
"	Seconds	um	Microns
rad	Radians	in	Inches
A	Angstrom	ft	Feet
mm	Millimeters	uin	Micro inch
cm	centimeters	mil	A thousandth of an inch
nm	nanometer		

Table 1.1

The arithmetic operators shown in Table 1.2 can be used.

Operator	Description	Example
+	Addition	2+3=5
−	Subtraction	3-2=1
*	Multiplication	3*2=6
/	Division	6/2=3
\	Integer division	9\2=4
^	Exponent	3^2=9
pi	Pi (3.14159265)	1*pi = 3.14159265
%	Percentage	25%=0.25
!	Factorial	5!=120 5!=5*4*3*2*1

Table 1.2

The math functions shown in Table 1.3 may be used.

Function	Description	Example
Abs( )	Absolute value	Abs(-7) = 7
Acos( )	Arc cosine	Acos(-1) = 3.141593
Asin( )	Arc sine	Asin(1) = 1.570796
Atan( )	Arc tangent	Atan(0) = 0
Cos( )	Cosine	Cos(3) = 0.99863

(Table 1.3 continued)

Function	Description	Example
Cot( )	Cotangent	Cot(1) = 0.642093
Csc( )	Cosecant	Csc(1) = 1.188395
Exp( )	Exponent (base e)	Exp(4) = 54.59815
Exp2( )	2 to the power of	Exp2(4) = 16
Exp10( )	10 to the power of	Exp10(3) = 1000
Log( )	Natural logarithm (base e)	Log(16) = 2.77259
Log2( )	Log base 2	Log2(8) = 3
Lo10( )	Log base 10	Log10(1000) = 3
MIN( )	Minimum value	MIN(1,2,4) = 1
MAX( )	Maximum value	MAX(2,8) = 8
Sec( )	Secant	Sec(0) = 1
Sin( )	Sine	Sin(30) = 0.5
Sqr( )	Square root	Sqr(9) = 3
Tan( )	Tangent	Tan(45) = 1

Table 1.3

Appendix 5: New Features of SolidWorks 2015

Following are some features that are new to the 2015 edition of SolidWorks software:

1) If you hover your mouse over a feature in the tree, you will now see the **Dynamic Reference Visualization** 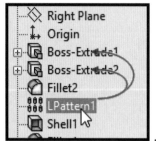. This shows parent/child relationships.

   To turn this feature on/off, right click on your part name at the top of the feature tree and then click on the icon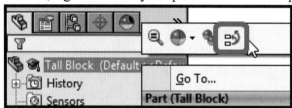

2) When adding a material, you now can click in the material window to open a Web browser and download more materials from the Web:

3) When you click or right click on a face in the graphics area, you see a new option to **Open Drawing**

4) There is a new line creation option, the **Midpoint Line**, which enables you to start your line creation from the midpoint rather than from an endpoint.

5) In the **View Heads up** tool bar, the **Zoom to Selection** icon is now included. This will zoom to the selected features.

6) The **Add-Ins** can be enabled/disabled from a new tab in the ribbon bar:

7) The **Add-Ins** window also displays the time required to load each add-in:

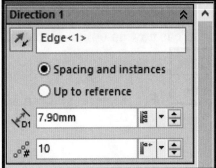

8) **SolidWorks MBD** is new to the ribbon bar. (This feature is for paperless drawings and is based on the realization that our old 2D orthographic projections no longer meet today's 3D design needs.) This truly paperless option allows dimensions to be added to a 3D part that will only be viewed on a computerized device:

9) In the **Linear Pattern Feature,** there is now an option to set the number of instances to **Up to reference**, enabling them to be equally spaced along the length of an entity.

10) When starting a new drawing sheet, there is an option to set up the sheet's **Zone Parameters**. Here you can specify the margin size and the number of zones on your drawing sheet.

11) In an assembly, you can now open a part in the orientation and position that it is currently viewed as:

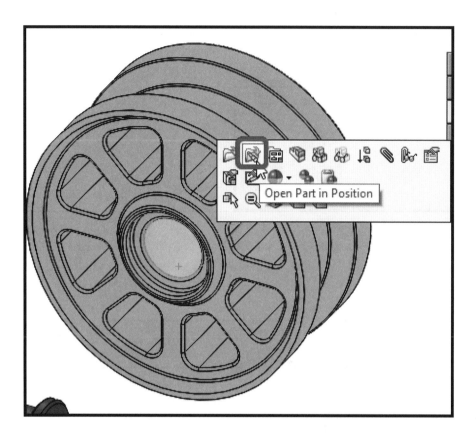

12) In an assembly, in the **Advanced Mates**, there is now an option to mate the center of one profile to the center of another . For instance, you can mate the center of a circle to the center of a rectangle.

Index

Add Sheet, 162
Add-Ins, 33, 203, 333
Advanced Mates, 286, 334
Aligned Section View, 184
Alignment, breaking and restoring, 176–177
Angle distance, 49–50
angled plane creation, 244–248
angles, dimensioning, 60–61, 79
Animation Wizard, 221–222
Annotation tab
   Balloon, 189
   Centerline, 17–18, 171, 236
   Display Annotations, 291
   Hole Callout, 225
   Magnetic Line, 189–191
   Model Items, 167, 177
   Note, 182
   restoring broken title block link, 163
   Revision Table, 195–197
   Tables, 188, 227–229
   See also Smart Dimension
Annotations, 5, 169, 170, 291
Appearances, 57, 95, 137, 139, 248
appendixes
   built-in calculator (mathematical functions), 329–331
   constraints/relationships, 325–326
   dimension tolerance types, 327
   short cuts, 328
   SolidWorks 2015, new features, 332–334
   arc sketching, 17–20
assemblies
   Assembly Configurations, 150–152
   bar puller, 204–218
   LEGO Minions, 127–134
   linear pattern, 141–144
   origin, 5
   suppressing components, 151–152
Assembly tab, 128, 187, 208, 296
axes, 7, 19, 65, 96–98, 127, 257, 264, 271

backing up files, 14–15
Balloon, 189
Bar Puller, 199–229
   adding a material, 203–204
   assembly and subassemblies, 204–218
   creating parts, 200–203
   drawings, 223–229
   exploded views, 218–222
   properties, 204
   Shank, 199–200
base (centering jig), 267–269, 289
bilateral dimension, 242–244
Bilateral Tolerance, 272–273, 327
BOM (Bill of Materials), 188–191
Boss/Base. *See* Extruded Boss/Base; Revolved Boss/Base; Swept Boss/Base
Boss-Extrude, 25, 26, 87, 89
Broken View, 223–224
Broken-Out Section, 181–182
built-in calculator, 329–331
bushings, 249–250, 283, 301–302

Calculate, 145, 316
calculator, built-in, 329–331
Cavity, 288–289
Center Block (centering jig), 270–273
Center Mark, 171–172
Center Point Arc, 77
Center Rectangle, 103, 113
centering jig, assembly, 283–307
   main assembly, 289–307
   screw subassembly, 283–289
centering jig, parts, 231–282
   base, 267–269
   bushings, 249–250
   Center Block, 270–273
   handle, 233–248
   handle grip, 232–233
   left jaw, 274–277, 279, 283, 287–289
   name plate, 280–282

centering jig, parts *(continued)*
    right jaw, 278–279, 283
    screw, 251–257, 283–289
    spring, 258–266
Centerline, 17–18, 171, 236
Chamfer, 49–50
Chamfer Dimension, 172–174
Change Transparency, 78, 140, 216
Circle (sketching tool), 27–28, 41
Circular Pattern, 55, 56–57, 81, 279
Circular Sketch Pattern, 73–74
Coincident mates, 132–133, 138, 206
Coincident relationship, 19, 21–22, 79, 90–91
Collapse Items, 132
Collinear relationship, 60
color, 57, 95–96, 137
commands
    adding to ribbon bar, 11–12
    customizing, 76
Comment, 194
components, 140, 150–152
Composite Curve, 258, 261
Concentric mates, 129–131
configurations
    animation, 221
    assemblies, 150–152, 284, 307
    axis creation, 96–98
    color, 95–96, 137
    Configuration Manager, 82, 86, 87, 110, 218, 221, 284
    derived, 99
    Design Table, 110
    design tables with Excel, 107–109
    Dimension, 85, 93–94, 265
    Exploded View, 218
    features, 94–104
    full round fillet, 104
    mark for drawing, 99–101
    Material, 266
    modifying, 94
    naming dimensions, 84
    parametric modeling, 82–104
    revolve cut, 96
    save as copy, 102–103

    selecting, 86
    showing part configurations, 191
    Toolbox, 33
Constant size fillets, 42–43
constraints, 19, 21, 87, 325–326
Contour Select Tool, 90
Convert Entities, 77–78, 100, 122, 253, 257, 321
Copy Settings Wizard, 35–36
Copy with Mates, 136–137, 218
Corner Rectangle, 37–38
Cosmetic Thread, 232–233
Create Parts, 33–34
Crop View, 224–225
cross sections, 53, 166–167
Curves, 7, 246, 254, 258–259, 261, 265
Customize
    commands, 76
    keyboard, 10–11
    Mouse Gestures, 13
    ribbon bar, 11–12
    templates, 29–36
Cut Extrude, 73, 80
Cut-Sweep Properties Manager, 256

Deep Freeze software, 14–15, 35–36
defined. *See* Fully Defined; Under Defined
degree minute seconds, 275–277
Depth field, 25
Derived Configuration, 99
derived parts, 273, 278
Derived Sketch, 259–260
design configurations
    design tables with Excel, 107–109
    naming dimensions, 84
    parametric modeling, 82–104
Design Library, 211–212, 218
Design Table, 107, 110–111
design tables with Excel, 105–111
Detail View, 183
diameter dimensioning, 62–64, 185, 252
dimensions and dimensioning
    angles, 60–61
    bilateral, 242–244
    Configure Dimension, 85, 93–94, 265

dimensions and dimensioning *(continued)*
   diameter, 62–64, 185, 252
   Dimension Property Manager, 169, 175
   Dimension Text, 172, 175, 185
   Dimension tolerance types, 327
   Dimension Type, 161
   Dimensions window, 107
   Dual Dimension, 175
   "Fixing Dimensions" (video), 168
   Hide/Show names, 7
   icons, 170
   inserting into view, 177–178
   inserting solid creation dimensions, 167–171
   interior features, 51
   leader diameter, 171–172
   Limit Dimension, 252–253, 279, 327
   modifying, 21, 27–28
   moving/copying, 168
   naming, 84
   neatening, 169
   ordinate, 226–227
   precision, 173
   Primary Value, 175
   to a virtual sharp, 185–187
   Tolerance/Precision, 175, 242, 249–250, 252–253, 327
   See also Smart Dimension
Display as Diameter, 171–172, 186, 252
Display as Radius, 186
Display States, 95, 161
Display Style, 53, 158, 161, 187
Distance option, 52, 265
Document Properties, 275
drawing sheets, reordering, 187
drawing templates, 31–32
Drawing View Property Manager, 158, 159, 161–162
Dual Dimension, 175
Dynamic Mirror Entities, 12, 76, 77
Dynamic Reference Visualization, 332

edge plane, 237–238
Edit Appearance, 95
Edit Component, 131
Edit Feature, 46, 121, 142
Edit In Context, 279, 293
Edit Material, 203
Edit Part, 131, 288, 296
Edit Sheet, 32
Edit Sheet Format, 31, 160, 163, 174
Edit Sketch, 23, 26, 73–74, 87
Edit Toolbox component, 214–218
editing your work, 26–28
End Support part, 303
Equal relationship, 22
Equal Spacing, 55, 73
Equations, 27, 265–266, 292–295
Evaluate tab, 141, 145–146
Excel, design tables with, 105–111
Exit Sketch command, 22
Exploded Assembly, 153–156
Exploded Line Sketch tool, 153
Exploded View, 159–161, 218–222
External References, 297
Extruded Boss/Base, 25, 37, 39, 40, 50, 100, 238, 296–297
Extruded Cut
   centering jig, 268, 272
   creating ribs, 123
   derived configurations, 99
   Hole Wizard, 242
   ribbon design, 321
   Through All, 79–80, 251–252
   tire tread extrusion, 76
   wheel hub, 67–68
   with line, 263–266
Extrusion
   creating ribs, 123
   handle, 24–25
   round, 40–46
   square base, 37–40
   to a face, 51

Face Fillet, 304–306
Face/Plane
   Mirror, 92
   Surface, 247
faces, 48, 51, 323
favorites, 266

Feature Fillet, 42–43
Feature Manager Design Tree
    Appearances, 57
    configuring, 94–104
    deleting features, 87
Feature Manager Design Tree
    Extruded Boss/Base, 24, 25, 37, 39, 40, 50, 100, 238, 296–297
    folders, 27
    introduction, 5
    LEGO Minions, drawing, 162–163
    re-ordering items, 46–50
    Right Midplane, 93
    Roll Back Bar, 119
    selecting a configuration, 86
    suppressing/unsuppressing features, 94
Feature Properties, 23, 24, 81
features
    renaming, 26, 81
    types, 1
Features tab
    Chamfer, 49–50
    editing, 46
    Extruded Cut, 68, 76, 80, 242
    Fillet, 42–43
    Linear Pattern, 43–46, 55, 118
    Mirror, 56
    Mirror Face/Plane, 92
    Reference Geometry, Plane, 52, 88
    Revolved Boss/Base, 58–59, 65
    Revolved Cut, 97, 257
    Rib, 53, 90, 116
    round extrusions, 40
    Shell command, 45–46, 47–49
    Sweep Cut, 256
    Swept Boss/Base, 262
Features to Mirror, 92, 124
Features to Pattern, 44–45, 46
File Explorer, adding a part through, 137–139
File menu
    Make Assembly from Part, 127, 205
    Make Drawing from Assembly, 157
    Make Drawing from Part, 223
    Pack and Go, 152–153
    Properties, 30–31, 163
    Save As, 102–103
    Save Sheet Format, 32
files, backing up, 14–15
files types, 14
Fillet
    creating ribs, 119
    editing, 49
    Face Fillet, 304–306
    Feature Fillet, 42–43
    Full Round Fillet, 104, 119
    global variables, 294
    re-ordering, 46
    Sketch Fillet, 70–72
Flexible Subassembly, 148–149, 290
folders, adding, 302
Free Drag, 128–129
Front Midplane, 88–89, 102, 150
Front Plane, 5, 6, 52, 58, 66, 132
Full Round Fillet, 104, 119
Fully Defined, 18–19, 21–24, 63, 316

Global Variables, 292–295

handle (centering jig), 233–248
    adding an appearance, 248
    adding raised text, 244–248
    creating a basic loft, 234–238
    Hole Wizard, 239–244
handle grip (centering jig), 232–233
heat sink, 113–126
Helix/Spiral, 254, 258–259, 261, 265
Hidden Lines Removed, 80, 166
Hidden Lines Visible, 90, 166
Hide/Show Items, 6, 7, 143–144, 257, 291
Hold Down, 297–299
Hole Callout, 225
Hole Series, 207–211, 290–291
Hole Specification Property Manager, 239, 241–242
Hole Table, 227–229
Hole Wizard, 239–244, 268, 299
Horizontal Ordinate Dimension, 226–227
Horizontal Relationship, 19, 50–51, 59, 90
hub. *See* wheel hub

Index

Insert
    Component, 128, 148–149, 206
    into New Part, 273
    Mate, 206
    menu, 107
    Molds, 288
    Paste window, 160
Interference Detection, 145–146
interior features, 50–58
    adding color, 57
    adding material type, 52
    Circular Pattern, 55
    constraining to midpoint, 50–51
    cross sections, 53
    extruding to face, 51
    Mirror, 56–57
    offset plane, 52
    Parallel plane creation, 55–57
    practice, 58
    ribs, 53–54
Internal Extrude, 89, 95, 146
Isometric View, 67, 115, 187

jaws (centering jig), 274–279, 283–289, 293–294, 299–301, 307

keyboard short cuts, 9–11
Keys, 305

left jaw (centering jig), 274–277, 279, 283, 287–289
LEGO Minions, assemblies, 127–156
    adding mates, 129–133
    adding parts, 137–139
    assemblies, 127–134
    component insertion, 128–129
    configurations, 150–152
    Copy with Mates, 136–137
    copying components, 135–136
    exploded assembly, 153–156
    Interference Detection, 145–146
    linear pattern, 141–144
    Minion creation, 134–152
    mirror assembly components, 139–141
    Open Part, 146–147
    Pack And Go, 152–153
    subassembly creation, 133–134
    subassembly insertion, 148–149
LEGO Minions, designing, 37–104
    design configurations, 82–104
    interior features, 50–58
LEGO Minions, designing
    parametric modeling, 82–104
    re-ordering items in feature tree, 46–50
    tire, 74–81
    2 x 2 block, 37–46
    wheel hub, 58–74
LEGO Minions, drawing, 157–191
    adding center marks/line, 171–172
    adding text to a dimension, 172
    Aligned Section View, 184
    BOM (Bill of Materials), 188–191
    Broken-Out Section, 181–182
    Chamfer Dimension, 172–174
    creating drawing views using model view, 187
    cross section, 166–167
    deleting a view, 165
    Detail View, 183
    Dimension Property Manager, 175
    dimensioning to a virtual sharp, 185–187
    Drawing View Property Manager, 161–162
    Exploded View, 159–161
    Feature Manager Design Tree, 162–163
    Inserting Dimensions into One View, 177–178
    inserting solid creation dimensions, 167–169
    Model Items Property Manager, 169–171
    Projected View, 176–177
    restoring broken title block link, 163
    Save As PDF, 191
    scale, 165
    showing hidden lines, 166
    showing part configurations, 191
    smart dimensioning, 178–181
    Standard 3 View, 164
    View Placement, 158–159

LEGO Minions projects
    assemblies, 127–156
    designing, 37–104
    drawing, 157-192
    revisions, 193–197
Lights, 7, 137
Limit Dimension, 252–253, 279, 327
limit distance mate, 286
Line tool, 19–20
Linear Component Pattern, 140, 142, 298
Linear Pattern, 43–47, 55, 91, 118, 121, 333
line/arc sketching, 17–20, 59
Link to external file, 292–294
Link to Property, 163, 182
Lock Rotation option, 130, 133, 213–214
loft, 234–238
logo, uploading, 31–32

Magnetic Line, 189–191
Make Assembly from Part, 127, 205
Make Drawing from Assembly, 157
Make Subassembly Flexible, 148–149, 290
Mark for Drawing, 99–101, 255
Material, 5, 52, 203–204, 266, 332
Mates
    adding, 129–133
    Advanced Mates, 286, 334
    Arm Assembly, 213
    Coincident, 132–133, 138, 206
Mates
    Concentric, 129–131
    Copy with Mates, 136–137
    Insert, 206
    limit distance, 286
    mating surfaces/Select Other, 135–136
    Parallel, 146
    Screw, 287–288
    Symmetric, 286–287
    Tangent mate, 138–139
    WheelEye folder, 130
mathematical functions, 329–331
mating surfaces, 135–136
Measure tool, 141, 285, 299, 323
Menu bar, docking, 9

Merge relation, 70
Midplane, 88–89, 92, 93, 102, 136, 150
Midpoint, 50–51, 87
Midpoint Line, 332
Minions. *See entries starting with* LEGO Minions
Mirror
    Arm Subassembly, 217
    assembly components, 139–141
    copying a sketch, 92
    creating ribs, 123–124
    Dynamic Mirror Entities, 76, 77
    Entities, 64
    Face/Plane, 92
    LEGO Minion, 56–57
Model Items Property Manager, 167, 169–171
Model View, 187
Motion Study, 221–222
mouse gestures, 13

name plate (centering jig), 280–282
naming
    dimensions, 84, 106
    Hide/Show names, 7
    renaming features, 24, 26, 81, 160
    Sheet Properties, 162
neatening dimensions, 169, 171
New Exploded View, 153, 218
New Part, 296
New View, 89–92, 313–314
Normal To view option
    copying a sketch, 90
    creating ribs, 115
    derived configurations, 100
    editing a sketch, 26
    Hole Wizard, 239, 242
    inserting sketch picture, 310
    interior features, 50
    mating surfaces, 135
    ribs, 53
    round extrusions, 40
    wheel hub, 67
notes, adding, 174, 182

Offset Entities, 68–69
Offset From Surface, 276
offset plane, 52
Open Drawing, 332
Open Part, 146, 148, 193
Options, customizing, 33
ordinate dimensioning, 226–227
Orientation, 89–92, 127, 161, 187, 313–314

Pack And Go, 152–153
Parallel mate, 146
Parallel plane creation, 55–57
parametric modeling, 82–104
parts
    derived, 273, 278
    features, 1
    in-context, 296–303
    origin, 5, 6
Pattern Axis, 278
PDF, saving as, 191
Pierce relationship, 91, 256
Pins, 305
planes
    angled plane creation, 244–248
    default, 5
    edge plane, 237–238
    editing, 88
    Front Plane, 5, 6, 52, 58, 66, 132
    Hide/Show, 6, 7
    Midplane, 88–89, 92, 93, 102, 136, 150
    offset plane, 52
    Parallel plane creation, 55–57
    Reference Geometry, 52, 55, 88, 96
    Right Plane, 5, 6, 56, 89, 113, 122
    sketch plane, switching, 66–67
    Top Plane, 5, 6, 37, 66, 113
    work plane creation, 317–322
Plastics, 52
Power Trim, 69–70, 78
Precision. See Tolerance/Precision
Primary Value, 84, 175
Projected View, 176–177

Quick Access ribbon menu, 3

raised text, 244–248
Rebuild, 86, 87
Rectangle, 37–38
Reference Configuration, 159, 161, 164
Reference Geometry
    Axis, 96, 264
    definition, 169
    icons, 170
    Plane, 52, 55, 88, 96, 113, 237–238, 244–248
reference planes. *See* planes
relationships, 59–60, 75, 325–326
renaming features, 24, 26, 81, 160
reordering drawing sheets, 187
reordering feature tree items, 46–50
repairing issues, 87–89
Replace Components, 150–151
Restore Settings option, 36
Reverse Section Direction option, 53
Revision Table, 195–197
Revolve Cut, 96, 97, 257, 271
Revolve feature, 66, 80, 309–310
Revolved Boss/Base, 58–59, 65
Rib feature, 53–54, 55, 56, 90, 116
ribbon, layered, 321–322
ribbon bars. *See* toolbars
right jaw (centering jig), 278–279, 283
Right Midplane, 92, 93, 136
Right Plane, 5, 6, 56, 89, 113, 122
Rollback Bar, 47, 49, 92, 119, 121, 279
Rotation Angle, 73
Rotation Point, 73
round extrusions, 40–47

Save As, 102–103, 191
Save Settings option, 35–36
Scale, 161, 162, 165, 281–282
screw (centering jig), 251–257, 283–289
Screw mate, 287–288
Section View, 53, 166–167, 184
Selected Contours, 65, 79, 80
Select Loop, 100
Select Midpoint, 50–51, 255

Select Other, 135–136, 328
Sensors folder, 5
settings, saving and restoring, 35–36
Shaded with Edges, 91, 158, 187
Shank, 199–200
sharing a sketch, 80–81
Sheet Format/Size, 157, 162
Sheet Properties, 162
Shell command, 45–46, 47–49
short cuts, 9–11, 328
Show Items. *See* Hide/Show Items
Side Face Set 2, 121
sketches and sketching
    basics, 17–27
    copying, 90–92
    Derived Sketch, 259–260
    editing, 23, 26–28, 73–74, 87, 320–321
    exiting, 22
    Fully Defined, 18–19, 21–24, 316
    introduction, 8
    line/arc, 17–20
    Mirror Entities, 64–66
    renaming, 24
    repairing errors, 319–322
    ribs, 115
    sharing, 80–81
    short cuts, 328
    sketch plane, 66–67
    Smart Dimension, 60–61
    thread profile, 255
Sketch Fillet, 70–72
Sketch ribbon bar, adding commands to, 11–12
Sketch Tools, 280–281, 310–312
Smart Dimension
    Chamfer Dimension, 172–174
    diameter dimensioning, 62
    dimensioning a drawing, 178–181
    dimensioning angles, 60–61, 79
    dimensioning to a virtual sharp, 185
    editing a sketch, 27–28
    fully defining the handle, 21
    ordinate dimensioning, 226–227
    round extrusions, 41
    short cut, 328
    square base extrusion, 38

Smart Fastener, 210, 290, 295
SolidWorks MBD, 333
SolidWorks 2015, new features, 332–334
$SolidWorksSetup.zip, 29–30
Spiral. *See* Helix/Spiral
Spline, 312–314, 315–316
springs, 258–266, 283
square base extrusion, 37–40
standard center drill, 105–109
Standard 3 View, 164
Stock Size, 228–229, 250
subassemblies, 133–134, 148–149
Summary Information, 31
Suppress, 94, 151–152
Surface/Face/Plane, 247
sweep cut thread (screw), 253–256
Swept Boss/Base, 258–262
Symmetric mate, 286
Symmetric relationship, 75

Tall Block tab, 82, 137, 165
Tangent mate, 138–139
Tangent relationship, 22
Task Pane, 2, 30
templates, customizing, 29–36
Temporary Axes, 7, 55, 257, 287
text, adding, 172, 244–248
Through All, 71, 79–80, 99, 251–252
tire, 74–81
    angular dimensioning using points, 79
    Center Point Arc, 77
    Convert Entities, 77–78
    customizing commands, 76
    Dynamic Mirror, 76
    re-naming features, 81
    sharing a sketch, 80–81
    Symmetric relationship, 75
    through all extrude cut, 79–80
    tread extrusion, 76
title block links, restoring, 163
Tolerance/Precision, 175, 242, 249–250, 252–253, 272–273
toolbars
    Copy Settings Wizard, 35
    docking Menu bar, 9

toolbars *(continued)*
  expert option, 8
  Hide/Show Items, 6, 7, 143–144, 257, 291
  introduction, 7–13
  Mouse Gestures, 13
  ribbon bar customization, 11–12, 76
  short-cut keys, 9–11
  Sketch Ribbon Bar, 8
  Use Large Buttons with Text, 8
  View Orientation, 12–13
  View (Heads-Up) tool bar, 7
Toolbox, 33–34, 203, 211–212, 214–218
Tools, 33
Top Midplane, 136
Top Plane, 5, 6, 37, 66, 113
transparency, changing, 78
Trim Entities, 69–70
turkey call, 309–323
  creating revolved part, 309–310
  creating spline, 312–314
  creating work planes, 317–322
  deleting a face, 323
  editing spline, 315–316
  inserting sketch picture, 310–312

Under Defined, 19, 63
  See also Fully Defined
Units, 275
Units Precision, 173, 242, 272–273
Up to Vertex Extrusion, 303
user interface, 2–6

Vertex Extrusion, 303
Vertical Relationship, 22, 51, 59, 90
View (Heads-Up) tool bar
  Hide/Show Items, 7, 257
  Orientation, 12–13, 313–314
  Section View, 53
  Temporary Axes, 55, 96

View Cube, 12–13
Zoom to Selection, 332
View Layout tab
  Break, 223–224
  Broken-out Section, 181
  Crop View, 224–225
  Detail View, 183
  Model View, 187
  Projected View, 176–178
  Section View, 166, 184
  Standard 3 View, 164
views
  deleting, 165
  Drawing View Property Manager, 158, 159, 161–162
  See also View Layout tab
Virtual Sharp, dimensioning to, 185–187

wheel hub, 58–74
  aligned section view, 184
  Circular Sketch Pattern/Edit Sketch, 73–74
  component insertion, 128
  diameter dimensioning, 62–64
  dimensioning angles, 60–61
  dimensioning to a virtual sharp, 186
  Extruded Cut, 67–68
  Offset Entities, 68–69
  relationships, 59–60, 70
  Revolved Boss/Base, 58–59, 65
  Sketch Fillet, 70–72
  sketch mirror, 64–66
  subassembly creation, 133
  switching the sketch plane, 66–67
  trimming entities, 69–70
Windows short-cut keys, 11
Wrap feature, 281

Zone Parameters, 333

About the Author

Fred Fulkerson is a graduate of the Faculty of Education, University of Western Ontario, and of the General Machining program at Conestoga College in Ontario. He is a Red Seal Certified General Machinist and is a certified Mastercam and SolidWorks instructor.

He has been teaching for more than fifteen years, the last ten for Conestoga College, where he is currently a professor in the Mechanical Engineering department. He also has had the privilege of representing his country as the Canadian Mechanical CAD Expert at the World Skills competition in Leipzig, Germany, in 2013.

Believing that we should help out when and where we can, Fred volunteers for Skills Ontario and Skills Canada and helps chair the Mechanical CAD competitions each year. One percent of all sales of this book will be donated to World Vision.

Fred enjoys canoeing, white water kayaking, archery, cycling, and squash. He lives just south of Cambridge, Ontario, with his wife, triplet teenage daughters, a Basset hound, two cats, two rabbits, and a pond of goldfish.

Any comments or suggestions about this book may be emailed to Fred at fredumacation@gmail.com.